Canto is an imprint offering a range of titles, classic and more recent, across a broad spectrum of subject areas and interests. History, literature, biography, archaeology, politics, religion, psychology, philosophy and science are all represented in Canto's specially selected list of titles, which now offers some of the best and most accessible of Cambridge publishing to a wider readership.

IN THE COMPANY OF ANIMALS

IN THE COMPANY
OF ANIMALS

A study of human–animal relationships

JAMES SERPELL

University of Pennsylvania

CAMBRIDGE
UNIVERSITY PRESS

Published by the Press Syndicate of the University of Cambridge
The Pitt Building, Trumpington Street, Cambridge CB2 1RP
40 West 20th Street, New York, NY 10011-4211, USA
10 Stamford Road, Oakleigh, Melbourne 3166, Australia

First published by Basil Blackwell 1986
Canto edition published by Cambridge University Press 1996

A catalogue record for this book is available from the British Library

Library of Congress cataloguing in publication data
Serpell, James, 1952–
In the company of animals: a study of human–animal relationships
/ James Serpell.
p.   cm.
Includes bibliographical references and index.
ISBN 0 521 57779 9 paperback
1. Pets – History. 2. Pets – Social aspects – History.
3. Pet owners – History. 4. Human–animal relationships – History.
1. Title.
SF411.5.S47 1996
304.2'7. – dc20   96–15185 CIP

ISBN 0 521 57779 9 paperback

Transferred to digital printing 2003

WD

To
Christopher Harold Serpell
(1910–1991)

Contents

Illustrations

Preface to the first edition

Human attitudes to animals often appear extraordinarily variable and arbitrary. Consider just two examples. In India the cow is sacred, and its slaughter and consumption are taboo. As a result, cows wander about and proliferate unmolested in a society where humans regularly die from lack of food. Conversely, in Europe and North America, where people enjoy exceptionally high living standards, cows are treated for the most part as walking milk bars or hamburgers on legs. The domestic dog, meanwhile, has become the western equivalent of the sacred cow. Dogs are cherished and nurtured as man's best friend, and the idea of killing and eating one is virtually unthinkable. Yet, throughout much of the Near East dogs are reviled as symbols of all that is filthy and degraded, while in China, Korea and the Philippines they are cooked and devoured with enthusiasm.

Anthropologists have spent decades discussing the origins and significance of these curious cultural differences, and the debate has gradually polarized into two opposing factions. One side argues that human attitudes to animals are fundamentally materialistic; that the eating of cow or pig or dog flesh is prohibited in certain societies because there are (or there were in the not-too-distant past) sound, economic reasons for such restrictions. The other side rejects the materialistic approach and maintains that particular animals are not eaten because they have acquired emotional or symbolic connotations that render their consumption socially and psychologically unacceptable. I do not disagree substantially with either point of view (indeed, as far as possible, I have avoided the whole subject of dietary taboos). Rather, I argue that in our relationships with animals – as with people – emotional and materialistic considerations are both important and, at the same time, frequently in conflict. The weird and wonderful ways in which human beings have sought to resolve this conflict provide the central theme of this book.

My approach to this topic has followed a circuitous and decidedly unorthodox route. Eleven years ago, as a recently fledged zoology graduate, I became interested in the western practice of keeping animals as pets. I began exploring what I thought was the relevant literature, only to find that the whole subject was virtually *terra incognita*. Almost nothing had been written about it and there had been few, if any, scientific studies. This struck me as odd for two reasons: firstly, because pet-keeping is evidently extremely widespread and popular and, secondly, because it is anomalous; it involves considerable emotional and financial expense, but does not appear to serve any obvious purpose. Coincidentally, I was not the only one to find the subject intriguing. At roughly the same time, a number of other individuals in Britain, France and the United States were also turning their attention to pets, and I soon found that I was part of a group consisting mainly of veterinarians, psychologists, psychiatrists, medical doctors, social workers and other zoologists, all of whom, in different ways, had become fascinated with the nature and implications of this puzzling and largely unexplained phenomenon. Eventually, I became involved in research into human–pet relationships, and it was during this period that I began to realize why the topic had, until then, been ignored so consistently.

Pet-keeping, I soon discovered, is a subject encircled by a great deal of prejudice and misunderstanding. The exact nature of these prejudices varies from person to person, but all of them essentially boil down to a vague notion that there is something strange, perverse or wasteful about displaying sentimental affection for animals. In this book, I explore these preconceived ideas about pets and pet-keeping in some detail. I attempt to show how and where they originated and why, for one reason or another, I believe they are misleading or unjustified. I have also tried to explain why this peculiarly intimate relationship between humans and non-humans tends to evoke unsympathetic reactions, and in doing so, I was eventually forced to some surprising and, to me, unexpected conclusions about our attitudes to animals in general.

The subject-matter of this book is wide ranging, and I have frequently wandered beyond the boundaries of my own limited area of expertise. However worthwhile the motives, this is invariably a hazardous undertaking for any scientist, and I must apologize in advance if my treatment of other disciplines appears superficial or misguided. While writing, I became acutely aware that almost every chapter could potentially be expanded into a book on its own but, since the topic is one about which comparatively little has been written, I decided to describe as much as possible of the surrounding terrain, as well as the details of my own particular patch. Above all, my intention has been to stimulate interest in a relatively unexplored landscape of human behaviour, and I hope that others will be encouraged to expand and improve upon this preliminary sketch.

Many people have helped me in various ways during the preparation and writing of this book, and I am indebted to them all. My special thanks, however, are due to Pat Bateson, John Sants, Robert and Olivia Temple, Christopher Serpell, Barry Keverne, Stephen Hugh-Jones, John Davey, Kim Pickin, Paul Martin, Phyllis Lee and Julie Feaver. To Julie, in particular, I owe my gratitude and lasting affection.

Preface to the Canto edition

In 1986, when the first edition of *In the Company of Animals* was published, the literature on the subject of human–animal interactions and relationships was not extensive. At the time, ethologists and behavioural ecologists typically focused their attentions exclusively on relationships among non-human animals, while psychologists and social scientists restricted themselves to the human domain. Apart from a small handful of anthropologists and historians, hardly anybody seemed to be interested in exploring the extraordinary variety of interactions and relationships between humans and other non-human species. As one reviewer put it, the whole subject was one that seemed to have 'fallen between the cracks of specializations'.[1] The preparation of this new edition has given me an opportunity to assess how far this situation has now changed.

The most obvious difference to those working in the field has been the enormous proliferation of published literature. Two new journals, *Anthrozoös* (published since 1987) and *Society & Animals* (since 1993), are dedicated exclusively to publishing scholarly articles on the subject of human–animal interactions, and there has been an exponential growth in the number of books. More importantly, perhaps, articles concerning aspects of human–animal interaction are increasingly finding their way into mainstream academic journals. It appears that the established disciplines are finally waking up to the research potential of this traditionally ignored dimension of their respective fields. Social medicine and psychoneuroimmunology, two research areas critical to understanding the potential therapeutic benefits of human–companion animal relationships, have also undergone tremendous expansion. In 1986, the evidence that people's social lives could have a direct impact on their physical health was limited and often speculative. Now it is overwhelming. Happily, none of the findings described in all of this new literature have altered the overall picture, although many have helped to provide depth, colour or detail to areas that were previously unfocused.

The growth in literature has been matched by a growth in public awareness of animal-related issues. Within the last ten to fifteen years, the so-called animal protection movement has expanded like a balloon, and its supporters are now better informed, and exercise far more social and political influence and authority than ever before. During the same period, animal welfare science has emerged as a large and respectable field of inquiry, the results of which are being put to good effect, not only in the design of more humane housing and husbandry systems for animals, but also by informing legislation and public policy. The number and popularity of pets has also increased, while the notion of pets being good for one's health – still a relatively novel concept in 1986 – is now much more widely and readily accepted.

Somewhat ironically, this latter trend may appear to challenge the relevance of some of the original arguments contained in the book. In the section entitled 'The case against pets', for example, I present a detailed – and some might argue, tendentious – defence of the practice of pet-keeping against what I view as a longstanding propensity to trivialize or denigrate affectionate animal–human relationships in our culture. This propensity I attribute to the fact that pets confront us with an egalitarian style of relating to animals which is morally at odds with our ruthless treatment of economically useful species. Ten years ago, two reviewers took me to task for this, claiming that it was logically absurd to propose that we, as a culture, denigrate pets or pet-keeping when 'more than half of us are in the supposedly disparaged category'.[2] Since pet-keepers have continued to multiply in the ensuing decade, and companion animals now enjoy higher status than ever before, it might appear to some readers that there is no longer a case against pets to be answered.

Needless to say, I thought long and hard about this potential criticism of the second edition, and was on the verge of trying to reconstruct the entire early part of the book, when my eye was caught by a feature article in a recent edition of *The New Republic*, a popular current affairs magazine. The article was entitled: 'The Power of Pets: America's Misplaced Obsession'.

In addition to trotting out most of the usual objections to pet-keeping – the extravagance, the wasted resources, its supposed tendency to pervert our affections – the author, a professor at Cornell University, concluded that the 'importance pets are acquiring in our lives and the anthropomorphic qualities with which we endow them may signal a troubling tendency to blur, as never before, the distinction between humans and animals'.[3] Quite so. But this tendency is troubling, he went on to explain, not because of the thorny moral questions it raises concerning our cavalier exploitation of most other species, but because it 'may lead, and often has led, to the worst forms of biologism, racism or naturalism. The Nazi *Lebensphilosophie*, to cite an egregious example . . . Whenever the radical heterogeneity between human and animals is erased, the door is open to brutally eugenicist arguments advanced under the guise of biological necessity or the authenticity of nature.'[4] In other words, keeping animals for companionship is dehumanizing. By indulging in it, we risk degenerating morally to the level of the Nazis whose pseudo-Darwinian ideas about *survival of the fittest* promoted the mass extermination of fellow human beings! Those wishing to follow the oddly contorted logic behind this argument should read the article for themselves. Suffice it to say that it was enough to convince me that my original approach to these issues remains appropriate, and that pet-keeping still needs its defenders, whether or not it is currently a majority activity.

So how, if at all, does this new edition improve on the previous one? First and foremost, I hope, in terms of accuracy.

The original book attempted to cover a vast area of ground, and it inevitably did so superficially. It was a sketch rather than a completed picture, and some of the information it contained was from doubtful sources and, occasionally, just plain wrong. Wherever I noted these mistakes, or they were brought to my attention by others, I have corrected them with the help of more appropriate reference material. The book is still a sketch but hopefully a more accurate representation of the facts than it was before. I have also revised obsolete figures and statistics, as well as drawing the reader's attention to newer or better source references to previously established facts or claims. I have not attempted a new synthesis of all the relevant literature published since 1986, and I freely confess to being selective when inserting new material. I have tried to depart from the original text as little as possible, only doing so when I felt that the picture had changed or would benefit from some local enhancement. Only Chapter 7, the one dealing with social medicine, has been substantially altered to reflect the recent expansion and consolidation of this rapidly developing field. My overall aim, as before, has been to provide a broad and readable overview of a topic which I, personally, find fascinating, and which I believe has important implications for all of us.

Finally, I wish to thank the people who have helped me *en route*, especially: Anthony Podberscek and Martin Jenkins for supplying facts; Nicholas Humphrey and Gordon Dunstone for correcting some of my errors; Paul Martin for re-opening my eyes to the delights of psychoneuroimmunology; Hal Herzog for his overall encouragement; Josie Dixon and Linda Bree for their patience; Elizabeth Jackson for her invaluable help, and Jacqui Bowman for her unfailing love and support.

Acknowledgements

I am grateful to all the authors and publishers who gave permission for me to reproduce extracts from their books, in particular: Yale University Press for extracts from R. Nash, *Wilderness and the American Mind* (© Roderick Nash 1982), and M. N. Cohen, *Health and the Rise of Civilization* (© M. N. Cohen 1989); Peter Singer and Random House UK Ltd for extracts from P. Singer, *Animal Liberation*; Methuen & Co. Ltd for material from Konrad Lorenz, *Man Meets Dog* (© 1953 by Konrad Lorenz; © 1983 by Deutscher Taschenbuch Verlag, Munich; © for the English translation 1994: Deutscher Taschenbuch Verlag, Munich. The title *Man Meets Dog* was republished in the United States in 1994 by Kodansha International, New York, Tokyo & London); the Estate of Peter Fleming and Random House UK Ltd for P. Fleming, *Brazilian Adventure*; Harcourt-Brace-Janovich for permission to quote from Ellen B. Basso, *The Kalapalo Indians of Central Brazil* (© 1973 by Holt, Rinehart & Winston Inc.); HarperCollins *Publishers* Ltd for J. Campbell, *The Way of the Animal Powers*, and S. Baxter, *Intensive Pig Production*; for Keith Thomas, *Man and the Natural World: Changing Attitudes in England 1500–1800* (© Keith Thomas 1983) and for Mary Midgley, *Animals and Why They Matter* (© Mary Midgley 1983) extracts reproduced by permission of Penguin Books Ltd; for E. E. Evans-Pritchard, *The Nuer*, by permission of Oxford University Press; and for C. Hole, *Witchcraft in England*, by permission of B. T. Batsford, London.

Every reasonable effort has been made to contact copyright

holders of texts quoted from in this book. The publisher and the author will be pleased to hear from any copyright holder whom they have been unable to contact.

PART I

A paradox

CHAPTER I

Of pigs and pets

All animals are equal but some animals are more equal
than others.

George Orwell, *Animal Farm*

Until the end of the last global Ice Age, around 12,000 years
ago, the human population of this planet derived all of its
food and raw materials from wild animals and plants. Anthropologists have coined the phrase 'hunting and gathering' to
describe this form of basic subsistence economy.[1] Typical
hunter–gatherers live in small, closely knit family groups of
fewer than fifty individuals. They are generally nomadic or
semi-nomadic, moving from place to place, and establishing
temporary camps, according to the dictates of the seasons and
the availability of game and other natural produce. A characteristic sexual division of labour exists within these groups.
The men do most of the hunting and butchering of game, and
they manufacture an ingenious assortment of stone, bone and
wood implements and weapons for this purpose. The women,
hampered to some extent by infants and young children, perform most of the gathering; foraging around the temporary
camps for the edible fruits, seeds, nuts, tubers and other plant
materials which form the staple part of the family diet.[2]

Under normal circumstances, the hunting and gathering
lifestyle is not especially arduous or uncomfortable. As long
as the population remains small, and as long as the groups
keep moving so as to avoid exhausting local resources, there is
generally sufficient food to go round, and adequate spare time
to engage in leisure activities. Existence is far from being a

3

romantic idyll – from time to time there are food shortages and starvation, and the usual assortment of lethal and debilitating accidents and diseases. Nevertheless, the relationship that exists between hunter–gatherers and the natural resources on which they depend is a remarkably balanced one; they kill and eat what they need to survive, but only rarely exceed the capacity of the environment to replenish the temporary deficits they create.[3] Human beings managed to live in this way, more or less unchanged, for over 90 per cent of their history on this planet.[4]

The end of the Ice Age brought this long period of economic and cultural stability to an abrupt conclusion. Within the space of a few thousand years, a socioeconomic revolution took place which overthrew the existing order and replaced it with something entirely new and unprecedented. This revolution was initiated by the domestication of plants and animals, and some authorities have described it as the most important and influential episode in the history of our species.[5]

Judging from the archaeological evidence, the first species to make this transition from a wild to a domestic state was the wolf (*Canis lupus*). Domestic wolves, the ancestors of the dog, first made their appearance among the prehistoric settlements of the Near East somewhere between 14,000 and 12,000 years ago. They were closely followed by domestic sheep and goats. Somewhat later, around 9,000 years ago, domestic cattle and pigs were also being farmed in parts of Asia. Horses, asses, camels, water buffalo and domestic fowl followed them and, around 3,000 to 4,000 years ago, the domestic cat emerged from wild obscurity in ancient Egypt.[6] The domestication and cultivation of plants coincided with the development of animal husbandry. Wheat, barley and various other cultivated plants appeared early on in Europe and Asia, while in the New World, maize, potatoes and beans were farmed, along with a different assortment of domestic animals, such as the llama, the alpaca, the turkey and the guinea pig.[7] By about 4000 years BP (before present) all of our most important domestic plants and animals were already permanent fixtures of human society.

The birth of agriculture and animal husbandry marked the beginning of the end of traditional hunting and gathering. By the time of Christ, farmers and livestock herders had already ousted the hunters from at least one-half of the inhabited earth. At the time of the discovery (or rediscovery) of the New World in the fifteenth century, they probably occupied only 15 per cent. Here and there, a few groups, such as the North American Indians, made valiant attempts to repel the invaders, but for their pains they were either obliterated or reduced to slavery and abject dependency. Now, only tiny, isolated populations of true hunter–gatherers remain, eking out an existence in some of the most marginal and least exploitable corners of the globe.[8] With the exception of the so-called 'whaling industry' and the sporting activities of the leisured rich, hunting as a way of life has almost vanished from the face of this planet.

The shift from hunting to farming also produced a fundamental change in human relationships with animals. Traditional hunters typically view the animals they hunt as their equals. They exercise no power over them, although they may hope to persuade the animal to be more easily captured by means of certain magical or religious practices. This essentially egalitarian relationship disappeared with the advent of domestication. The domestic animal is dependent for survival on its human owner. The human becomes the overlord and master, the animals his servants and slaves.[9] By definition, domestic animals are subservient to the will of humanity and, for the majority of species involved, this loss of independence had some fairly devastating long-term consequences.

Take, for example, the unfortunate case of the domestic pig. The Eurasian wild boar (*Sus scrofa*), from which the pig is descended, is an intelligent, sociable mammal which is still surprisingly common in parts of Europe and Asia. It is a creature of open forests and woodland. The typical wild boar social group consists of a matriarchal herd or 'sounder' containing, perhaps, half a dozen closely related females and their offspring. Sub-adult males sometimes form bachelor herds, while mature males are generally solitary. Herds are not exclusively territorial, although they may be aggressive toward outsiders, and they generally roam over an area of about twenty-five hectares. Wild boars are most active during the daytime and around dusk, and spend much of their time foraging for food. They are omnivorous and feed on a wide variety of plants, including fungi, ferns, leaves, roots, bulbs and fruit, as well as on insects, insect larvae and earthworms, and small vertebrates such as mice and frogs. Much of this food is obtained by rooting around in the soil with the bony and muscular snout. Foraging parties are noisy, maintaining a continual, conversational exchange of grunts, squeals and chirps. At night these animals sleep *en masse* in large dens or nests. They are also extremely partial to wallowing in mud, an activity which helps to keep them cool in warm weather and rids the skin of external parasites. They are naturally clean animals, and deposit their excrement in specific dunging areas.

In the Northern Hemisphere, mating takes place in the autumn. The boar courts the sow by displaying and 'chanting' and by nudging her with his snout. He also champs his jaws together to produce salivary foam. The boar possesses lip glands which secrete a sexual scent or pheromone, the smell of which is highly stimulating to the sow. The production of salivary foam probably helps to disperse this pheromone. The boar also tests the receptivity of the sow by placing his chin repeatedly on her rump. If the female is receptive, she will 'stand' for the boar to permit mounting and copulation.

Farrowing takes place the following spring. As she approaches term, the sow leaves the herd and constructs a large nest of twigs, leaf litter and dried grasses in which she gives birth to up to twelve piglets. The piglets remain in the nest for about ten days before following their mother on her foraging expeditions and, eventually, rejoining the matriarchal sounder. At this age the piglets are playful and intensely curious. Although they have an exaggerated reputation for ferocity, wild boars obtained as piglets are easily tamed and make charming pets, almost dog-like in their affection and loyalty.[10]

The life of the modern domestic pig stands in sharp contrast to that of its wild progenitor. In the West, methods of farming pigs have been revolutionized within the past hundred or so years, and the trend has been toward increasing intensification. Gone are the days of the humble farmyard pig, contentedly rooting in the soil for various edibles or foraging in the woods for beechmast. Gone are the days of the mud-wallow or the appetizing bucket of swill, the leavings of the farmer's kitchen and other surplus farm produce. The pig of modern agribusiness is born and raised in artificial confinement throughout its brief and uncomfortable life. It has been reduced to the status of a strictly utilitarian object; a thing for producing meat and bacon.[11]

From the moment of conception, the intensively farmed domestic pig is regulated and controlled, and rarely permitted to engage in any of the natural activities enjoyed by its wild ancestors. About a week from giving birth, the sow is herded into a farrowing crate; a narrow steel cage in which she is able to stand up or lie down but is impeded from making any other movements. Despite this, sows engage in various stereotyped activities which have been interpreted as frustrated attempts at nest-building. These are accompanied by clear signs of distress. The use of these crates is justified by the desire to reduce piglet mortality, particularly by preventing sows lying or stepping on their young. In reality, the differences in piglet mortality between crates and other, more open farrowing systems may be quite small. Between one-tenth and one-third of a piglet per litter is the estimated improvement using crates.[12] The newborn piglets are allowed to suckle from their incarcerated mother for anything from a few hours to several weeks, depending on the rearing methods employed. (A short period of suckling is essential in order to give the piglets the opportunity to acquire passive immunity to certain diseases from the mother's milk and colostrum.) Under less intensive systems, the piglets remain with the mother in small weaning pens until they are anything from three to seven weeks of age. The pens are equipped with 'pig-creeps', the porcine equivalent of cat-flaps, which allow the piglets access to their own separate trough of food.[13] Modern textbooks on pig-rearing, however, recommend removing the piglets from the mother as soon as possible – within twelve to thirty-six hours after birth. This has several advantages to the producer. Sows deprived of their piglets stop lactating and become sexually receptive again more quickly. Early removal also reduces the likelihood of the piglets acquiring some infection from the filthy conditions surrounding the farrowing crate.[14] It is customary to carry out a list of routine operations on the piglets as soon as possible after they are born: their 'needle teeth' or incisors are clipped, their tails are docked, their ears are notched for identification, and the males are castrated. No anaesthesia is employed during these operations.[15]

In the most intensive systems the piglets are generally isolated within hours of birth in small individual cages which are stacked, row upon row, in tiers. Nourishment is supplied in the form of regular, controlled doses of liquid food at roughly hourly intervals. To get an idea of the totally impersonal and technological nature of this pig-rearing process, it is worth quoting verbatim from a major textbook on the subject:

In most refined systems, each piglet is contained within its own isolated space and provided with its own air-conditioned environ-ment. There is no direct contact between piglet environments and the surrounding room environment. Slightly less refined is the provision of a crypto-climate for the group of piglets, with aerial connections between the piglet spaces and the surrounding room environment. Each piglet space is, however, supplied with a thermal conditioning element in the form of an overhead heating element or a heated floor pad. A third system provides only for a controlled room environment, with the individual piglet places constructed of open wire mesh cages.[16]

The author's chief objection to the latter system is that farm personnel, working in the room, may become uncomfortably hot.

At seven to fourteen days, the piglets are moved again to new quarters where they are housed in groups in slightly larger cages. In these cramped and boring conditions the

young animals are inclined to engage in what are euphemistically termed 'social vices'; chiefly biting or sucking each
other's navels, tails and ears, apparently out of sheer
frustration. To combat this behaviour, it is recommended that
the piglets be kept hot (22 to 27°C) and therefore lethargic, in
near darkness, and free from sudden disturbances.[17] Pigs
reared in these artificially confined conditions are delicate and
notoriously susceptible to stress. Sudden noises or bright
lights make them frightened and potentially aggressive
towards each other. They may also induce a condition known
as PSS or 'porcine stress syndrome', an affliction characterized
by extreme stress, rigidity, blotchy skin, panting, anxiety and,
often, sudden death.* PSS can strike factory-farmed pigs at
any age, but it is particularly galling to the producer when the
pigs are close to market weight after several months' investment of food.[18]

Once onto solid food, the weanling piglets or 'weaners' are
grown on in small groups in pens until they reach slaughtering
weight at around six to eight months of age. For ease of
cleaning, the pens have concrete or slatted metal floors, and
no bedding is provided. It is obviously difficult to assess
whether pigs are comfortable on such floors, but all the
evidence is to the contrary. Given the choice, pigs prefer to
stand or lie on sand or straw bedding, and foot deformities and
lameness are common in animals raised on hard floors without
access to softer bedding areas. Most pigs are slaughtered
before serious deformities have time to develop, so there is
little economic incentive to farmers to provide more comfortable conditions. Aggression and social vices are prevalent in
the unpleasant and overcrowded conditions of the growing
and finishing pens so, once again, the pigs are kept in total or
partial darkness for most of the time.[19]

Finally, once they reach a suitable size and weight, the pigs
are subjected to the terrifying ordeal of transportation
and slaughter. One day, after months of inactivity, boredom

---

* It is now established that certain genetic strains of pig bearing the 'halothane' gene
are more susceptible to PSS (Grandin, 1994).

and frustration, they are herded out of their pens and crammed, like so many sardines, into a livestock truck where they spend hours or even days, virtually unable to move and without food or water. Those which are understandably frightened and uncooperative are not dealt with lightly. Handlers are generally in a hurry and they are frequently provoked to the point of undue violence, usually administered through the toe of a boot, a stick or a club or, more often nowadays, through the tip of an electric prod. Pigs are maimed, bruised and killed during transport and, in the words of Peter Singer:

Animals that die in transit do not die easy deaths. They freeze to death in winter and collapse from thirst and heat exhaustion in summer. They die, lying unattended in stockyards, from injuries sustained in falling off a slippery loading ramp. They suffocate when other animals pile on top of them in an overcrowded, badly loaded truck. They die from thirst or starve when careless stockmen forget to give them water or food. And they die from the sheer stress of the whole terrifying experience, for which nothing in their lives has given them the slightest preparation.[20]

At the abattoir, some pigs exhibit every symptom of abject terror; screaming and jostling one another in a nightmare of blind panic.[21] Ideally, death is relatively quick and painless, the animal being stunned by an electric current or a captive-bolt pistol before its throat is cut. Unfortunately, the circumstances of slaughtering are not always so humane.* If stunning is performed inexpertly, the animal probably suffers more than it would from having its throat cut. One author describes the situation in a nutshell: when slaughtering is 'done well by caring people, pain and misery can be minimized; done badly, untold horrors will be routine'.[22]

The pigs that go to slaughter are, arguably, the lucky ones. A few unfortunate sows and even fewer boars may be selected for breeding. Nowadays, to discourage aggression, breeding

---

* Grandin (1988: 205) reported witnessing 'deliberate acts of cruelty occurring on a regular basis' at 32 per cent of the slaughter plants she surveyed in the United States.

sows are generally kept isolated in individual narrow stalls in which they are unable to turn round.* They remain in these stalls until they come into oestrus or 'on heat', an event which is often detected by sitting on the sow's back or rump when, like the wild boar sow, she will exhibit the 'standing' reflex.[23] As soon as she signals her condition in this way, she is herded quickly out of her pen and into one of the boar pens where, ideally, nature is allowed to take its course. More often, if the sow is frightened or shows signs of being coy and uncooperative, the mating procedure becomes nasty and brutish. The sow is pushed, shoved and physically restrained by stockmen in order to get her into a suitable position for mounting. The boar is likewise discouraged from engaging in any of the normal preliminaries of courtship and, because artificial selection has so altered the physical proportions of the domestic pig, he is often unable to achieve intromission without manual assistance. Once a successful copulation has been performed, the boar is pushed aside, and the sow is summarily herded back to her stall where, more often than not, she remains until farrowing.[24]

Needless to say, research has found methods of streamlining this clumsy and time-consuming business of copulation and conception. Modern techniques of artificial insemination can dispense with 'natural' mating altogether, although, once again, the sows are not always obliging. Like their wild ancestors, domestic sows find the sight, sound and smell of a sexually active boar stimulating, and they are more inclined to conceive in the presence of one. Undeterred by this, research has yielded an aerosol can containing artificial boar pheromones with which the sow can be sprayed to get her in the mood. Some artificial insemination programmes have even resorted to playing tape-recordings of boar mating cries to their sows. Hormone injections are also widely used to

---

* During the last ten years, pressure has grown within many European countries for the adoption of humane alternatives to the stall-housing of dry (i.e. pre-oestrous or gestating) sows. UK legislation has recently phased out the practice of tethering sows in stalls, a procedure which further restricted their ability to move.

accelerate and control the sow's reproductive activities. To reduce the amount of time between conceptions, and to ensure that farrowing takes place in convenient batches over short periods, hormones can be injected into the sows while they are still suckling their piglets. Hormones are also used to induce sows to give birth at a time of day which suits the farmer.[25] In short, pig-breeding is concerned with just two aims: to increase the rate of piglet production, and to reduce the non-productive periods of the sow – the periods during which she is consuming valuable food, but is not actually gestating or suckling piglets. As one textbook bluntly puts it: 'The sow has one commercial purpose in life which is to produce weaners and the more efficiently she does this, the higher will be the profit margin on any pig enterprise.'[26]

Nobody in their right mind would consider the existence of the modern domestic pig a pleasant one. But for those who engage in factory farming or benefit from its produce, this kind of callous and brutal treatment is easily justified. Like many other species on this planet, humans like to eat meat, and they are prepared to kill and inflict a certain amount of suffering on other animals in order to indulge this preference. If we assume, for the sake of argument, that people have a right to go on eating meat, as their ancestors have done for about three million years, then it obviously makes sense for us to exploit our domestic livestock in the most cost-effective and efficient way. And this is precisely what modern, intensive farming is all about. The actual methods employed may vary from species to species, but the basic principle remains the same: maximize productivity; minimize costs.[27] From start to finish, modern agribusiness is based on this simple industrial formula. The pig breeder aims to produce the largest number of weaners per sow per annum, the grower seeks to get his pigs to slaughtering weight in the shortest possible time, the transporter wants the animals loaded and delivered with the minimum delays, and the slaughterman is chiefly concerned with increasing the rate at which he can kill and butcher the pigs that arrive at his abattoir. And at the end of the chain stands the spectre of the voracious consumer, who is solely interested in buying the highest quality meat and bacon for the lowest possible price. The fact that this principle also ensures that the livestock involved are subjected to a lifetime of continual deprivation, distress and discomfort seems to be largely irrelevant; merely an unfortunate by-product of the harsh, economic necessities of life. And the minority of people who display genuine moral concern for the welfare of farm animals often seem to be regarded as either stupid, sentimental or just plain crazy.

This kind of hard-nosed, economic attitude to the exploitation of domestic animals is a simple and straightforward one, and it is one that is tacitly endorsed by the majority of people in the western world. Humans have a right to eat meat; farmers have a duty to supply this demand as cheaply as possible; animals inevitably suffer as a consequence. Why complicate the issue with imponderable questions about the morality of it all? Indeed, as a basic philosophy it would be exceedingly difficult to fault, if only it were consistent; if only it applied right across the board to all our dealings with all domestic species. But, quite clearly, it does not. There exists in our society an entirely separate category of domestic animals which, for no obvious reason, is exempt from this sort of treatment. These animals are, of course, the ones we normally refer to as pets.

According to 1994 figures, an estimated 36 million pet dogs and 35 million pet cats currently live within the countries of the European Union, together with a further 173 million other miscellaneous pet animals, ranging from cage birds, rabbits and guinea pigs through to reptiles and aquarium fish. Over half of all the households in the EU contain at least one such animal, and a substantial minority contain several.[28] The figures for the United States are even more remarkable: 54 million dogs, 59 million cats, 16 million birds, 7.3 million reptiles and amphibians, and 12 million fish tanks distributed among some 56 per cent of all households.[29] Most of these animals belong to domesticated species, just like our old friend the pig. Yet few of them serve any significant practical purpose. We do not slaughter and eat them. We do not milk them or scramble their eggs. We make no use of their fur or their hide, and we do not harness them to ploughs. In economic terms the majority are completely useless. Yet we allow them the run of our houses, give them personal names, and treat them as honorary members of the family. We stroke them, cuddle them, play with them, groom them and ensure that they receive all the exercise and social contact they need to keep them happy and healthy. They are regularly supplied with specially prepared, vitamin-enriched food, provided with warm and comfortable places to sleep, and at the first signs of illness, are immediately taken to expensive and highly trained doctors. And when they eventually expire, they are mourned like departed loved ones, even to the extent of being buried with full ceremonial honours.[30]

Pets, particularly dogs and cats, can also be a considerable source of embarrassment and inconvenience to their owners. They limit a person's freedom and independence; they may be noisy, dirty, smelly or disobedient and, in some cases, they may exhibit behaviour problems such as aggression, anxiety, destructiveness and hypersexuality, all of which can turn them into a serious nuisance. Pet-ownership is undoubtedly a major responsibility.[31] Yet it seems to be a responsibility which most owners are prepared to take on, despite all the potential drawbacks.

The financial costs of pet-keeping are equally staggering. Americans spend around $8.5 billion annually on dog and cat food, and the most recent figures published by the American Veterinary Medical Association show that they also spend in excess of $7 billion on veterinary care for pet dogs, cats and birds.[32] In Britain, the figures are scarcely more modest. British pet-owners spend in the region of £1.3 billion ($2 billion US) each year on dog and cat food and treats.[33] Other expenses are largely a matter of guesswork, but if we allow for a conservative doubling of costs in the last decade, then it is likely that roughly £300 million is spent on various pet accessories, such as cat-litter, treats, medication and cosmetics, aquaria and equipment, leads, collars, clothing, toiletries, beds, toys, etc., while a further £350 million goes on veterinary fees and drugs.[34] And these are only the direct expenses. A whole host of indirect costs also arise from pet-keeping, the main burden of which is currently shouldered by taxpayers. For example, British government surveys suggest that dogs are causal factors in about 0.6 per cent of all road accidents involving injuries to people. Admittedly, the vast majority (76.4 per cent) of these injuries were slight but, nevertheless, a substantial proportion were classified as serious, and a few (0.9 per cent) were fatal. Other surveys indicate that dogs are also responsible for causing up to 16 per cent of road accidents not involving injuries. A 1984 study estimated the total financial cost of these accidents in terms of lost output, damaged property, health care, police work and administration at around £40 million per annum. More recent figures are not available but clearly one would need to double or even treble this amount to allow for inflation over the last decade.

Animal bites and scratches form another indirect cost of pet-ownership. It has been estimated that between 100,000 and 200,000 people require medical treatment for dog bites annually in Britain,[35] and the average cost to the health service for each treatment is about £70. This would represent a potential cost to the taxpayer of up to £14 million per year. In the United States, dog bite has been described recently as a problem 'of epidemic proportions, affecting more than 1% of the US population annually and accounting for widespread exposure to zoonotic diseases and more than 20 fatalities each year'.[36] Even the most conservative estimates suggest that close to 600,000 people a year require medical treatment for dog bite in the USA[37] and, in these days of escalating health costs, the total bill for treating all the victims is unlikely to be less than $120 million. Dogs also cause serious and fatal injuries to domestic livestock. It is difficult to estimate the total cost in compensation to farmers, but roughly 10,000 animals in Britain, chiefly sheep, poultry and cattle, are killed or injured by dogs every year.[38] As carriers and transmitters of infectious diseases, pets also represent a significant health hazard, although the extent of the danger to humans has been grossly exaggerated by the anti-pet lobby.[39] Of the few disease organisms transmitted from pets to people, the majority cause relatively minor ailments such as local infections from bites and scratches, diarrhoea, and skin rashes. A small number are more serious. Aside from rabies, the disease which has attracted most attention in recent years is toxocariasis, a condition caused by the larvae of two kinds of nematode worm, *Toxocara canis* and *T. cati*. These worms normally inhabit the digestive tracts of dogs and cats, respectively. Humans can become infected with *Toxocara* through contact with soil contaminated with animal faeces, or through direct contact with infected pets. The most exaggerated national estimates suggest that roughly 16,000 people, mainly children, are infected by *Toxocara* every year in Britain, although the vast majority never develop any clinical symptoms and probably suffer no permanent harm. In a small proportion of cases (about 10 per year in England and Wales), the larvae migrate through the bloodstream and may lodge in the liver, the brain, the lungs, or in the back of the eye where they cause, respectively, liver-enlargement, epilepsy, asthma-like respiratory problems, and impaired vision.[40]

Canine and feline waste products are not only a potential source of infection, they can also be a considerable nuisance. On an average day the dog population of Britain deposits 4.5 million litres of urine and 1 million kilograms of faeces, some of it in parks and other public places where it is aesthetically objectionable, and where it can interfere with human recreation. It is impossible to put a price on the nuisance value of this kind of environmental pollution. Nevertheless, it is undoubtedly a source of disgust and inconvenience to many people.[41]

Figures and statistics of this kind will doubtless raise a small forest of eyebrows but they are not intended, in this case, to arouse anti-pet feeling. They are provided merely to illustrate the formidable investment which pet-keeping represents; an investment which, it seems, pet-owners and pet-owning societies are prepared to make despite significant costs and in the absence of any measurable economic return. In other words, as far as pets are concerned, the simple rules which normally govern our treatment of domestic animals no longer apply. Instead of maximizing productivity, minimizing costs, and turning a blind eye to the welfare of the animals involved, we do exactly the opposite. The economic benefits of pet-keeping are negligible at best. Yet the majority of pet-owners spare no expense to ensure that their animals are as happy, contented and secure as they possibly can be.

Of course, pets have had to pay a price for the pampered status they enjoy. Dogs and cats and other pets are not allowed the sort of freedom they would experience as wild animals. But it is doubtful whether most of them suffer much as a consequence. A well-fed cat is a fairly inactive creature, and even those which are confined all their lives to small flats or apartments seem to experience little distress. Dogs are such intensely sociable animals that most of them willingly remain close to their owners rather than wander at will. Purely as a matter of convenience, the majority of cats and a substantial number of dogs are neutered at an early age. These operations generally deprive pets of normal sexual interest, and obviously prevent them from having offspring or participating in the pleasures of parenthood. But again, the degree of suffering caused by this infringement is probably minimal. Surgery is performed under anaesthetic, and it is unlikely that the animal has any concept of the possible pleasures it might have enjoyed were it still physically intact. Artificial selective breeding has transformed the dog into a bizarre variety of shapes, sizes and temperaments, and not all of these changes have been in the animal's best interests. Many breeds suffer from congenital physical disorders, some of which condemn them to a lifetime of discomfort.[42] Pets are also subjected to unnecessary and inhumane 'cosmetic' surgical procedures, such as tail-docking, ear-cropping or declawing,* either to suit

---

* In Britain, the vast majority of veterinarians have long been ethically opposed in principle to tail-docking, ear-cropping and the surgical 'declawing' of cats for non-medical reasons. Until recently, however, many were willing to perform tail-docking operations in order to prevent lay persons from carrying out these procedures inexpertly. In 1993, despite opposition from the UK Kennel Club, it

the owner's convenience, or simply to make them conform to the arbitrary standards imposed by the animal-breeding and showing community. Animal abuse and cruelty to pets also remains a serious problem in many western countries, and not all pet-owners are the responsible, caring people they ought to be. Although estimates vary, it is likely that at least half a million pets are lost, disowned or abandoned by their owners in Britain each year. A substantial proportion of these are never reclaimed or rehomed and are therefore humanely destroyed in animal shelters.[43] In the United States the situation is worse. According to the results of the best regional estimates, nearly 6 million American dogs and cats finish their lives in animal shelters each year.[44] But such abuses pale into insignificance when compared to the sort of treatment which is meted out on a regular basis to domestic pigs, poultry and veal calves.

When beagles or cats are used in laboratory research, militant animal liberationists are provoked to acts of mindless violence. Imagine the level of public outrage if it were disclosed that the government was subsidizing the factory farming of puppies and kittens. Indeed, the recent story of 'The man who had to eat his dog to survive' occupied the entire centre page of one of Britain's major newspapers.[45] The man who had to eat his pig to survive would scarcely qualify for a footnote. Yet in China, Korea, the Philippines and several other areas of the world, the slaughtering and eating of dogs and cats is relatively commonplace.[46] So why should we treat these animals any differently? Are dogs and cats somehow more worthy of moral consideration than pigs or chickens? Do they possess some intrinsic quality that entitles them to favoured treatment? Dogs and cats are, arguably, more aesthetically appealing than pigs. But so what? The singer, Madonna, is more pleasing to look at than the Pope but this

became illegal in the UK for a non-veterinarian to dock a puppy's tail, a move which has brought about an effective ban on cosmetic tail-docking. In addition, the European Commission has drafted a proposal to prohibit all cosmetic surgical procedures throughout the EU (Hubrecht, 1995). All of these procedures are still commonplace in the United States and Canada.

hardly entitles her to a position of moral superiority. Pigs are no less intelligent than dogs or cats; they are sociable and clean and, when tamed, make amiable pets.* So why are they treated so badly? Why are they not given the same ethical consideration as cats and dogs?

Here, then, is the paradox. At one extreme are the animals we call pets. They make little or no practical or economic contribution to human society, yet we nurture and care for them like our own kith and kin, and display outrage and disgust when they are subjected to ill-treatment. At the other, we have animals like the pig on which a major section of our economy depends, supremely useful animals in every respect. The modern pig is one of the most efficient converters of feedstuffs and domestic and agricultural waste into edible protein. For every 3.5 kilograms of food it consumes it gains a kilogram in weight, and almost every part of its anatomy is of material or nutritive value.[47] We pickle its trotters, make black puddings from its blood, sausages from its intestines, and expensive and durable leathergoods from its skin. We even emulsify its thick white fat for the production of ersatz ice-cream. And in return for this outstanding contribution, we treat pigs like worthless objects devoid of feelings and sensations. By rights, we ought to be eternally grateful to pigs for the extraordinary sacrifices they make on our behalf. Instead, the quality of life we impose on them suggests nothing but contempt and hatred.

The reason this state of affairs is so disquieting is that we seem to be simultaneously harbouring two totally contradictory and incompatible sets of moral values. Indeed, the situation would be morally and psychologically intolerable if we gave both sets of values equal weight. But we do not. The

---

* Since 1985, when a Canadian farmer imported the first 19 specimens, Vietnamese pot-bellied pigs have become the latest fashionable pet in Europe and North America. According to a 1991 estimate, 46,000 pigs, ranging in value from $50 to $10,000, are kept as pets in the United States (Yost, 1991). It would be interesting if this new pet-keeping trend exerted a positive influence on attitudes to swine in general, although the most recent press reports suggest otherwise. Animal shelters in England now appear to be inundated with unwanted pot-bellies (*Sunday Times*, 27 August 1995).

inconsistencies inherent in our treatment of these two separate classes of domestic animal are only paradoxical when it is assumed that both types of treatment are normal. The contradictions are swiftly resolved by viewing either one extreme or the other as odd, unusual or exceptional and, therefore, unimportant in the overall scheme of things. I will argue that, until very recently, we have accepted the least painful solution to this disturbing moral dilemma. Instead of questioning the hardline, economic exploitation of animals we have tended, in one way or another, to adopt a disparaging, condescending or trivializing attitude to pets and pet-keeping.

PART II

The case against pets

CHAPTER 2

Substitutes for people

What is a dog anyway? Simply an antidote for an
inferiority complex.

W. C. Fields, attrib.

At first glance, it might appear supremely illogical to claim
that the practice of pet-keeping tends to be denigrated by our
society when a majority of householders go out of their way to
engage in this activity, and presumably enjoy doing so. Nevertheless, the claim is readily justified. Firstly, pet owners have
only very recently become a majority in the West, and a
slender majority at that. Twenty years ago, there were half as
many pet dogs and cats in the United States as there are
today, and this appears to be a continuation of a trend that
goes back at least as far as the Second World War, and
probably much further. Historically, pet-owners have always
been the minority, and their current proliferation is certainly
unprecedented. Attitudes to pet animals, such as dogs and
cats, have also changed. Although this is obviously difficult to
document, various historical studies have shown that strong
affection for animal companions did not become widespread in
Europe until the nineteenth century.[1] Moreover, American
veterinarians have noted a dramatic increase in the average
level of their clients' emotional attachments for pets during
just the last twenty to thirty years.[2]

Secondly, in spite of the current explosion of pet-keeping
in the western world, and despite the vast emotional and
financial investment which all these animals represent, there
have been few serious attempts to explain either why

pet-keeping exists or what purpose, if any, it serves. In fact, only within the past two decades has the subject attracted any significant scientific interest at all, and the little research which this interest has generated has barely succeeded in scratching the surface of the problem. As the psychologist Nicholas Humphrey noted in 1983, the institution of pet-keeping 'has been given astonishingly little attention by social scientists. No one seems to have noticed that in the United States there are nearly as many cats and dogs as there are televisions. The effects of television have been minutely researched and documented, but the effects of pets still remain virtually unanalysed.'[3] This persistent failure to pursue the possible causes and functions of pet-keeping is hard to explain, except as a symptom of prejudice.

Although we still know little about the nature of this phenomenon, nearly everybody cherishes his or her own particular theory on the subject, and most of these preconceived ideas tend, in one way or another, to denigrate pet-keeping to the point where it is not considered important or respectable as an area of scientific enquiry. The most widespread and popular of these theories – and the one which has undoubtedly done most to trivialize the whole topic – is the belief, or at least suspicion, that pets are no more than substitutes for so-called 'normal' human relationships. The Roman writer Plutarch was among the first in a long succession of pet-haters to put this suspicion into words:

Caesar once, seeing some wealthy strangers at Rome, carrying up and down with them in their arms and bosoms young puppy dogs and monkeys, embracing and making much of them, had occasion not unnaturally to ask whether the women in their country were not used to bear children; by that prince-like reprimand gravely reflecting upon persons who spend and lavish upon brute beasts that affection and kindness which nature has implanted in us to be bestowed on those of our own kind.[4]

This view of pet-keeping as a 'gratuitous perversion'[5] of natural behaviour has been reiterated time and again throughout history and, nowadays, is most often expressed either by means of caricatures of post-menopausal women and

poodles or by a general tendency to regard people's relationships with their animal companions as absurd, sentimental
and somewhat pathetic. As the psychiatrist Aaron Katcher
points out, 'we are taught to despise the sentimental, to think
of it as banal or as a cover for darker hidden emotions'.[6] In the
case of pets, these darker, more sinister aspects are seen as
taking a variety of forms. Some appear to believe that pet-
owners are somehow socially inadequate and that they use
their pets, in much the same way that drug-users use heroin,
as artificial and ultimately detrimental substitutes for reality.
Others view the relationship as merely an excuse for playful
domination.[7] Others refer to its 'essentially sexual nature'[8]
and still others regard it as a thoroughly pernicious phenomenon that consumes people's positive emotions and, thus,
contributes indirectly to 'the oppression and physical or
psychological annihilation of human beings'.[9] Unfortunately,
for those who hold such beliefs, there is no shortage of strange
and sensational stories and case histories to confirm their
worst suspicions. Indeed, it is worth playing devil's advocate
for a moment simply to illustrate just how easy it is to obtain
a distorted impression of pet-keeping.

The popular and tabloid press devotes almost as much
coverage to human–pet relationships as it does to people's sex
lives, although not all of these accounts are necessarily critical
of the practice.* Press stories about pets and their owners fall
into two basic categories: those aimed at pet-lovers which
stress the almost supernatural loyalty and devotion of the
animal for its master – the dog that rescues the drowning child
or the cat that pursues its owner across hundreds of miles of
trackless wilderness – and those which emphasize the extraordinary extremes to which pet-owners will go for the sake of

---

* Another peculiar characteristic of almost all journalistic reports of pets and pet-
keeping is the compulsive urge to pun. As Richard Klein (1995: 19) has recently
observed, the subject of pets seems to 'invite a form of linguistic playfulness, a sign
that the tone is being dropped down to a level more appropriate to the topic'.
Although Klein would probably disagree, I would argue that this journalistic habit
is merely symptomatic of a universal tendency to trivialize the whole phenomenon
of pet-keeping.

their animals.[10] Sometimes the latter involve tragedy – people dying while attempting to rescue their pets from burning houses, stormy seas or frozen rivers – but more often they are designed simply to amuse, fascinate or horrify. It is mostly these that tend to promote a jaded view of pet-keeping. Examples are legion. Take, for instance, the following headline from a popular Fleet Street journal:

WILL THEY LET LUCY LIVE?
THE QUESTION THAT CLOUDS THE LIFE OF A
FINANCE MINISTER

This particular story dominated the British press for several days in January 1984, and concerned no less a personage than Albert Gudmundsson, Iceland's Minister for Finance, who was quoted as saying that he would rather resign his job and go into exile than allow the authorities to do away with his elderly mongrel dog, Lucy. Over the years the worthy Mr Gudmundsson, together with about 3,000 other Icelanders, had been discreetly flouting a sixty-year-old local law which specifically bans dogs from the city of Reykjavik. Unfortunately, a radio disc-jockey with a grudge informed the police about the unlucky finance minister. The Chief of Police then had no choice but to insist that the dog either be destroyed or banished from the city. The incident caused a tremendous furore, with Mr Gudmundsson publicly proclaiming that Lucy was a beloved member of his family, that he would never give her up, and that he would fight to the last ditch to have the law abolished. A contemporary survey revealed that the majority of Icelanders supported the Chief of Police.[11]

Of course, it could be argued that Mr Gudmundsson was putting up less of a fight for his dog than for a point of principle, but then what about the strange case of Marmaduke (Gingerbits), alias 'Sonny', the cat that made English legal history in March of the same year? The trouble all began when police constable John Sewell and his wife went on holiday leaving behind their ginger cat Marmaduke, formerly Marmaduke Gingerbits (the name was suitably abbreviated following the cat's operation), to be looked after by a friend.

According to the friend, Marmaduke disappeared soon after they went away. Meanwhile a neighbour of the Sewells began receiving visits from a large, well-fed ginger cat, and one evening she observed another neighbour, Mr Monty Cohen, trying to catch the animal and claiming that it was his missing pet, Sonny. Despite the animal's reluctance to be caught, she accepted the story and let Mr Cohen take the cat home. Only later, when she read the Sewells' advertisement offering a £10 reward for the return of their beloved Marmaduke did she realize her mistake and inform the Sewells. PC Sewell then went round to Mr Cohen's, recognized his cat and demanded him back. Mr Cohen refused and a scuffle ensued in which Mr Cohen was put in an arm-lock and PC Sewell was punched and eventually evicted catless. The matter was then taken to the civil court. Nine months and £6,000 in legal costs later (the cat was officially valued at £5), the judge finally ruled that 'Exhibit B' was indeed the Sewells' cat Marmaduke, and that Mr Cohen had made an understandable mistake. Sewell was ordered to pay Cohen £50 for trespass and £200 toward his legal costs. Cohen was ordered to pay all of Sewell's costs and was reported to be 'visiting his bank manager'. The British taxpayer shouldered the rest of the bill, and the national press had a field-day.[12]

Another recent case involved a French student who was so devoted to her pet rat that she risked severe penalties by smuggling the animal into Britain concealed in her pullover. In the event the trip proved disastrous. The unfortunate rat was mistaken for vermin and was stamped on by the patrons of a London pub. The girl was so distressed by the murder of her pet that she took a drug overdose and, when she eventually recovered from this, she was promptly fined £400 for violating quarantine restrictions. And all for a rat. Or the stranger than fiction story of the Texas hairdresser who was somewhat upset when her pet Rottweiler killed and ate her four-week-old daughter, but who wept hysterically when told that the animal would have to be destroyed. 'I can always have another baby', she said after the incident, 'but I can't replace my dog Byron.'[13]

Baby-eating Rottweilers aside, pet-owners are quite often prepared to make costly and painful sacrifices for the sake of their animals. Victoria Voith, an American veterinarian who specializes in treating behaviour problems in pets describes some fairly typical case histories. The man, for example, whose two-year-old Weimaraner hound systematically destroyed, chewed and sometimes devoured his furniture and household effects whenever he left it alone. During consultation the man admitted that the dog had cost him roughly $3,000 in damages but said that he was too fond of the animal to consider parting with it. Or the couple whose pet Dobermann began displaying signs of aggression toward their eighteen-month-old son and eventually bit the child severely on the face for no obvious reason. The owners were advised either to give the dog away or to have it destroyed. Despite repeated warnings about the danger of further attacks on the child, the couple still had the dog a month later. Even more bizarre was the case of the young woman who for several years had tolerated the company of an elderly poodle that bit her on the arms and back while she was asleep. The dog shared the bed with her at night because it refused to sleep on the floor and barked incessantly if restrained or shut out of the room. The woman had never mentioned the animal's disturbing habits to her vet because she was afraid that it would have to be put down.[14]

But, just to even the score, it is not always owners who are the victims of mentally disturbed animals. One Californian animal behaviour consultant mentions a divorced and childless woman who induced an ulcerated colon in her pet German shepherd and reduced it to a nervous wreck by spoon-feeding it while it was forced to sit on a chair at the dinner table.

America seems to provide particularly fertile ground for the growth of extreme animal–human relationships. New York companies, such as 'City Dogs' and 'Canine Styles', sell dog-size, leather biker and bomber jackets, ski parkas, sweaters, T-shirts, panties, Spandex doggy swimsuits, tuxedoes, sequined evening gowns, and matching rain ponchos for dogs and their fashion-conscious owners.[15] Los Angeles pet boutiques and superstores supply over-indulgent owners with custom-made water beds, gold-plated choke-chains and personalized (or animalized) leather-covered dining suites for the pet who has everything. For outdoor dogs they offer leather backpacks for camping and hiking, taffeta bow-ties for the well-dressed dog-about-town, and real or imitation-mink stoles for bitches of quality. Not to mention a host of more mundane items: Burberry raincoats for the wet month of February, pullovers, frilly frocks, underwear and canine cosmetics such as nail varnish, available in a wide choice of fashionable colours. Some boutiques will even make all the arrangements for a pet's birthday party – embossed invitations and special catering, including enormous meat-loaf 'birthday cakes' in the shape of fire-hydrants, etc.[16] Most of these accessories are aimed at devoted dog-owners, although some pet couturiers are prepared to design clothing for parrots or even hamsters. One of the latest luxury items for budgerigars (parakeets) consists of a warm shower which automatically switches itself on when the bird perches in the appropriate position. Cedar-chip, mint-scented and pulverized corn-husk cat-litters are some recent innovations for the benefit of pampered pussies, while fans of the movie *Jurassic Park* may also wish to invest in a four-foot-high, dinosaur-shaped scratching post costing $130.[17]

Hard-working, upwardly mobile American pet-owners can enroll their animals in daytime playgroups (although the waiting lists for membership are reputedly enormous), and when they go on vacation they are not averse to using establishments such as the Pet Set Inn, 'One of America's Finer Hotels for Dogs' where guests enjoy the freedom of fully carpeted, temperature-controlled apartments, each with piped 'new wave' music and its own private sun porch.[18] And if the hotels are all booked up, there are always summer camps. According to its advertisement, Campo-Lindo For Dogs in upstate New York offers 'spacious accommodations and a program of recreation for your canine camper. Each dog enjoys his own private cabin with 400 square feet of romping space, choice of meal plan and lots of activities, ranging from your basic fetch and catch to swimming.'[19]

Standards of medical care for pets are also rapidly approaching those enjoyed by people. So called 'high-tech' veterinary procedures, such as magnetic resonance imaging (MRI), hip replacements, transplantation surgery, heart pacemaker implantation, cataract surgery, radiation therapy and chemotherapy for cancer, and various advanced emergency and intensive care procedures are becoming increasingly *de rigueur* in teaching hospitals and specialist veterinary clinics across Europe and North America. And such treatments are by no means cheap. In the United States, a standard course of chemotherapy for a 30 kilogram dog can easily run to $2,000, while a total hip replacement may cost the owner as much as $2,500. There is, apparently, no shortage of demand for these types of procedures, a fact which generates a modicum of controversy in a nation where many humans are unable to afford adequate medical care.[20]

Even with the best medical attention in the world, no pet is immortal. But American pets at least have the consolation of a wide choice of tasteful final resting places. Pet cemeteries are big business in the USA, offering their customers all the facilities of a human burial ground. For example, the 'Pet Haven' in Gardena, California, a four-and-a-half-acre site containing the remains of more than 28,000 animals including the favourite dogs of Edward G. Robinson and Nat 'King' Cole, provides the owners of the deceased with a cemetery plot; a flower container; the services of a 'groomer'; a choice of caskets (satin-lined redwood or water-proof polyurethane), and viewing rooms containing plastic flowers and effigies of the Virgin Mary where owners can hold services or merely commune with the Dear Departed. Black diamond-granite headstones etched with portraits of the deceased, and perpetual 'memory lights' are optional extras.[21]

Since the eighteenth century, English aristocrats and men of letters have commemorated the deaths of favourite pets with monumental sculptures and florid epitaphs. Occasionally these were fanciful arrangements – mere excuses to display wit or to satirize contemporaries – but more often they betrayed genuine emotional trauma on the part of the owner.[22] Lord Byron's inscription and epitaph written in remembrance of his dog, Boatswain, not only sings the animal's praises but also, in the last few lines, clearly expresses Byron's personal sense of loss:

> Ye! who perchance behold this simple urn,
> Pass on – it honours none you wish to mourn:
> To mark a friend's remains these stones arise;
> I never knew but one, – and here he lies.[23]

A certain amount of grief or remorse when a pet dies is only natural, especially if the animal has been an important part of its owner's life for many years. But, occasionally, mourning over the death of an animal companion can become pathological. The psychiatrist, Kenneth Keddie, has documented specific case histories: that of a sixteen-year-old schoolgirl, for instance, who became depressed and developed a livid rash on the skin of her hands within twenty-four hours of the death of her thirteen-year-old King Charles spaniel. The family doctor prescribed a course of antihistamines. The girl experienced initial difficulties swallowing the tablets and, within forty-eight hours, was unable to swallow anything at all, including food and water. The girl was admitted to a psychiatric hospital where her condition improved as soon as she was allowed to discuss the death of the animal with a therapist. In another case, a married woman of fifty-five was brought into a psychiatric outpatient clinic suffering from a long list of depressive symptoms including listlessness, insomnia, weight loss and anorexia. It eventually transpired that these symptoms dated from the death of her fourteen-year-old pet poodle eighteen months before. She admitted to being very dependent on the dog and to treating it like a child. When it died she had been unable to bring herself to remove its basket or bone and spent most of her waking hours thinking about the animal.[24] Favourite dogs and cats seem to be the most frequent objects of pathological grief reactions. A case reported in the British press, for example, concerned a middle-aged man who composed a four-page memorial letter to his dead Pomeranian dog and then hanged himself. But the death of almost any pet can produce profound and lasting distress, particularly in children.[25]

Pathological mourning over a pet is itself often the inevitable outcome of pathological over-dependency. Again, some striking case histories are found in the psychiatric literature. In one case a middle-aged woman developed a mildly disfiguring facial disease and became chronically self-conscious about it. Over a period of months she gradually withdrew from social contact with people and became increasingly dependent on her pet dog. For seven years she lived as a virtual recluse, devoting herself entirely to the animal which she never let out of her sight. When the dog died she developed an intense and prolonged grief response for which she was forced to seek psychiatric attention. An even more extreme case involved a thirty-two-year-old divorcee whose husband had left her because of her inability to love him as much as she loved her dog. Despite a distinctly ambivalent and destructive relationship with her mother, she moved into a neighbouring house and the two women then established a curious triadic relationship in which the dog became the focus of mutual involvement. One day, after a bitter argument, the woman forcibly ejected her mother from the house, killed the dog and then committed suicide. The mother later admitted that their final quarrel had arisen when the mother demanded total custody of the dog because the daughter 'wasn't loving him enough'.[26]

As George Bernard Shaw once lamented, pet animals sometimes 'bear more than their natural burden of human love'. When this happens, it tends to be the animals who suffer. So-called 'animal collectors' – people with a compulsion to adopt stray animals in such numbers that they eventually overwhelm the person's ability to provide them with adequate care – are an increasingly common problem in Europe and North America. According to one recent report, 200–300 cases are reported each year in the United States alone, and these are generally only the most extreme examples. Most of them follow a similar pattern: the person (or persons) involved has a modest income, is relatively isolated, and is often the victim of a recent emotional loss or bereavement. He, or more usually, she, becomes obsessed with the desire to save stray animals from untimely deaths, either on the streets or at the hands of humane societies, and typically there is an extreme reluctance to part with any animal, even to a good home, once it has been adopted. As the animals gradually accumulate, sooner or later their numbers exceed the resources and facilities available, and neighbours then begin to complain. By the time public health authorities are able to intervene, the animals are frequently so malnourished and diseased that immediate euthanasia is the only humane option.[27]

Excessive devotion to pets acquires a particularly unsavoury quality when it is combined with misanthropic or xenophobic attitudes to other people. According to reliable accounts, Adolf Hitler and several other prominent Nazis, including Göring, Goebbels, Hess and Admiral Dönitz were all extremely attached to pet animals, especially dogs. Hitler's extraordinary affection for dogs was noted long before he rose to power, and during the last weeks of the war he became so emotionally dependent on his last dog, Blondi, that he risked his own life by taking her for a walk each day outside his bunker.[28]

People's interactions with their pets, particularly dogs and cats, characteristically involve relatively large amounts of uninhibited physical contact and intimacy. It is not altogether surprising, therefore, that this physical and sensual dimension of the human–pet relationship occasionally overflows accepted boundaries. Zoophilia and bestiality – sexual relations between people and animals – are probably as old as animal domestication. Nevertheless, wherever and whenever they occur, they are generally hedged about by powerful taboos and mystical or totemic connotations. In many parts of South East Asia, Australasia and North America the original indigenous peoples trace their ancestry back to primordial sexual encounters between women and dogs (rarely men and bitches).[29] The centaurs, satyrs and minotaurs which populated the myths of ancient Greece bear similar testament to man's early zoophilic fantasies and practices. In Greece, zoophilia acquired magical or religious expression in the amorous adventures of Zeus and other gods who periodically transformed themselves into a variety of animals in order to consort with female mortals. Such mythical seduction scenes were occasionally enacted on stage in sequences which sometimes included actual coitus between animals and humans.[30]

Other religious creeds were less liberal. Bestiality was specifically prohibited in the Hittite code, the Old Testament and the Talmud, and under Christianity it was regarded as one of the most heinous and unspeakable of all crimes. In medieval Europe, men and sows, men and cattle, and men and donkeys were publicly tried and burned together for acts of sodomy, and at Toulouse a woman was burned for having intercourse with a dog.[31] In Britain, bestiality became a capital offence in 1534 and, apart from one brief hiatus, remained so until 1861. In 1679 a woman and a dog were hanged together for the offence on Tyburn Hill outside London.[32]

Probably because of the moral outrage and disgust which it traditionally evokes in our culture, it is difficult to obtain good evidence for the prevalence of zoophilia and bestiality in contemporary western society. In his famous 'reports' Alfred Kinsey and his associates found cases in about 8 per cent of the American male population, although the incidence rose to between 17 and 50 per cent (depending on one's definition of bestiality) among adolescent boys living on farms or ranches. Only 3.6 per cent of adolescent women reported equivalent behaviour – most of it with dogs – and in only 1.2 per cent of these encounters was there repeated genital contact to orgasm. The reports also emphasized that, 'no other type of sexual activity . . . accounts for a smaller proportion of the total outlet of the total population for both males and females'.[33] In other words, bestiality seems to be extremely rare and most of what occurs is between rural male adolescents and livestock of one form or another. Only very occasionally does it arise as an extension of the emotional bond between an owner and his or her pet. Having said this, however, it is clear that human–pet relationships involving an erotic or sexual component do occur from time to time and, once again, there are some fairly bizarre stories in the literature to confirm this.[34] Alan Beck and Aaron Katcher, in their book *Between Pets and People*, mention one extraordinary case of a female psychiatric patient, suffering from a persistent psychosomatic disorder, who agreed to be interviewed in front of an entire class of students. During the interview she casually revealed, among other things, that she had regular sexual intercourse with the family dog.[35]

It is obvious from these peculiar anecdotes and case histories that *some* pet-owners develop feelings and attachments for their pets which equal, and occasionally exceed, those they have for other people. They are prepared to make costly, humiliating or even heroic sacrifices on their behalf, they dress them up and indulge them sometimes to the point of cruelty, and they may be crippled with remorse when they die. And, while the animal is still alive, they may become emotionally dependent on it or even engage in sex with it. It is perhaps understandable that many people should regard such conduct as either the product of emotional inadequacy or as a curious perversion of normal social or sexual behaviour, but it must be emphasized that all of these relationships represent extremes of human–animal intimacy. They tell us little, if anything, about the unremarkable majority of human–pet relationships; the kind which exist in roughly half the homes in Europe and North America; the kind that never make it to the front pages. After all, millions of ordinary people drink alcohol, but we do not, as a rule, regard them as ridiculous, unbalanced or depraved simply because a small minority become alcoholics. It is hardly fair, then, to view the entire practice of pet-keeping as unnatural or pernicious merely because a few of its adherents exceed the limits of acceptability. In order to justify such a view one would need hard evidence, first, that pet-owners possess flaws in their personalities which predispose them to seek the companionship of animals rather than people and, second, that close relationships with animals somehow detract from or interfere with people's ability or desire to relate to other human beings. Fortunately for pet-owners, recent research in this area offers very little support for either of these suppositions.

Attempts have been made in recent years to discover differences in personality or 'psychological status' between pet-owners and non-owners. Unfortunately, the evidence which has accumulated presents a somewhat complicated and confusing picture. One of the earliest findings emerged quite fortuitously in 1956 from a study which set out to explore methods of detecting so-called 'authoritarian' personalities. One of the items in the questionnaire employed in this study was the statement, 'dogs are much more admirable animals than cats'. It later transpired that strong agreement with this statement was much more strongly associated with authoritarianism than the replies to any other item in the questionnaire. The author of this study argued, reasonably enough, that dogs are naturally more obedient and subservient than cats and are therefore strongly preferred by authoritarian people.[36] The preceding story about Hitler and Blondi immediately springs to mind, but the author is quick to emphasize that his discovery, while valid for whole populations, does not apply reliably to individuals.

Far more damning accusations against pet-ownership came from two later studies published in 1966 and 1972, respectively, in which it was claimed (a) that pet-owners did not like people as much as non-owners, (b) that they did not feel as liked by others, (c) that they liked their pets more than they liked people, and (d) that urban pet-owners tended to have weak egos.[37] Unfortunately, or perhaps fortunately (depending on one's viewpoint), the instrument employed for assessing relative liking for people and pets was distinctly crude, and the statistical methods used were highly questionable. Notwithstanding these shortcomings, the authors made the outspoken assertion that 'pet-owners are less psychologically healthy than non-owners' and that 'psychologically the pet seems to function as a detriment to effective social relationships and consequently to the person's mental health'. Since neither of these conclusions was warranted from the data, it is reasonable to assume that they tell us more about the personal prejudices of the researchers than about the nature of pet-ownership.

It is only fair to set against these findings, and the story of Hitler and Blondi, the fact that many of history's foremost humanitarians were animal lovers and, often, pet-owners. The English poet and moralist, Alexander Pope, wrote poems about his pets, and the eighteenth-century moral philosopher, Jeremy Bentham, who supported the abolition of slavery, was one of those 'unfortunates who shared an equal liking for cats and for mice'.[38] Another prominent abolitionist, William Wilberforce, joined the committee of the newly formed Society for the Prevention of Cruelty to Animals, and a later Secretary of the Society, John Colam, was instrumental in founding the National Society for the Prevention of Cruelty to Children. Richard Martin who pioneered the first Protection of Animals Act, also championed the abolition of the undue use of the death penalty, while Henry Bergh, the founder of the New York SPCA in 1866, was also responsible for the first successful legal prosecution against child abuse in the famous case of 'Little Mary Ellen'.[39] More recently, Albert Schweitzer, who wrote a book about his pet pelican, and devoted much of his life to the care of people afflicted with leprosy, advocated a philosophy of 'Reverence for Life', while Mahatma Gandhi argued that the greatness of a nation could be judged by the way it treated its animals.[40]

Notwithstanding these worthy examples, the idea that pet-keeping can be detrimental in certain situations has received a modicum of support from the clinical findings of some psychiatrists. In a 1984 paper, one of them refers to what he calls the 'pet trap' or the tendency for some people to become so enmeshed and involved emotionally with their pets that they lose interest in other people. This constitutes a hazard because, once 'trapped' by the pet, the person may feel no further motivation either to alter their social circumstances or to seek psychiatric help.[41] The point is a valid one but, again, it probably only applies to exceptional individuals.

The case histories provided by the author as evidence all involved people who either had pre-existing psychological problems or who had become, through no fault of their own, isolated or alienated from more appropriate human partnerships; in other words, people with no alternative outlets for social interaction. It is doubtful whether individuals lacking these kinds of relationship problems would ever be in danger of being 'trapped' by their pets in the same way.

On a more positive note, psychologists at the University of Oklahoma found that people who cherish affectionate attitudes toward dogs also tend to feel the same way about people. They invited 200 university students to complete questionnaires which measured their liking or affection for dogs and, on the basis of their replies, grouped them into Low Affection, Moderate Affection and High Affection categories. A fortnight later, an equal proportion of students were randomly selected from each category, and all of them were asked to complete another test which measured their ability to establish, and their desire for, affectionate relationships with other people. At the time the students had no idea that the two questionnaires were in any way related. From the results it emerged that students in the Moderate Affection category were significantly more affectionately disposed toward people than those in the Low Affection group. In addition, men reporting little affection for dogs apparently desired relatively little emotional involvement with other people. The only sour note in these findings was the discovery that High Affection students were less affectionate toward people than Moderates. With commendable circumspection, the authors suggest that this is 'not inconsistent' with the popular belief that fanatical animal lovers are people who have displaced their affection from people to pets.[42]

A related but unpublished study in England obtained comparable results in 1976. In this case, 50 dog-owning and 100 non-owning Yorkshire householders were compared for their 'desired affiliation' for other people – i.e. their overall desire for close relationships and friendships. The author distinguished between owners who spent a lot of time interacting with their pets and those that maintained more passive relationships, and he found that the interactive owners (but not the passive ones) expressed a significantly stronger desire for affiliation than the non-pet-owners. This, it was concluded, disproved the contention that 'dogs substitute for human relationships to the point of exclusion, and supports the view that a pet may be owned by persons who are unable to satisfy their affiliative needs'.[43]

In yet another study, this time in California in 1981, elderly pet-owners and their non-owning counterparts were compared using an entirely different set of psychological instruments. The authors concluded from their results that pet-owners were significantly more self-sufficient, dependable, helpful, optimistic and self-confident than non-owners, although they were quite rightly unwilling to say whether they thought these differences were causes or effects of pet-ownership.[44]

Finally, there have been a couple of studies whose findings are more or less neutral. In one case, psychologists measured 'well-being' and 'self-acceptance' in pet-owners and non-owners using a well-known test called the California Psychological Inventory. At first it appeared from the results that male non-owners were scoring more highly on the well-being scale than male pet-owners. But on closer inspection it was noticed that half the non-owners belonged to social clubs as against only a quarter of pet-owners. They re-ran the test removing club-members and the apparent differences in well-being promptly disappeared.[45] In another more recent study, researchers in Brooklyn looked for differences in personality within a large sample of university students who were grouped into current pet-owners, former owners and non-owners. The personality dimensions measured included anxiety, depression, mood, Type A (or coronary prone) behaviour, and sensation seeking. No significant differences could be found between the three groups on any dimension which could not be accounted for by the influence of some other intervening factor, such as where the subjects lived or the type of accommodation they occupied.[46]

The idea that pet-owners, on average, may be more or less the same as everyone else is also borne out by indirect evidence. For instance, if pets really function as surrogates and their owners are genuinely using them simply to replace close relationships with people, then one would expect to find pet-owning at higher frequencies among single persons – bachelors, spinsters, widows, etc. – i.e. among the people most likely to suffer from feelings of social alienation and loneliness. Indeed, the popular stereotype of the little old lady with the pampered Pekingese reveals how common such associations are generally assumed to be. In fact, no clear-cut relationship of this kind exists. The pet-food industry and veterinary profession have carried out exhaustive surveys on the distribution of pet-ownership among different social and economic groups within society and, contrary to expectations, most of these surveys have found higher frequencies of pet-ownership among couples, families with children, and in large households than they have among single or elderly people.[47] This result may, in part, be due to differences in lifestyle or accommodation type. Dog-ownership in particular is known to be more common among people living in detached houses (presumably with substantial gardens) than it is among those occupying terraced housing or flats. In Japan where floor space is at a premium, the correlation between home size and pet-ownership is particularly strong. The percentage of Japanese owning dogs is about 4.4 in homes of less than 60 square metres. But it rises to nearly 30 per cent among those enjoying more than 100 square metres of floor space. The statistics for other pets are somewhat different. Bird- and fish-ownership is, predictably, less dependent on accommodation size, and there appears to be room to swing a cat in any Japanese home with a floor space greater than 60 square metres.[48] All of which helps to illustrate the dangers of generalizing about the kinds of factors which may or may not predispose people to become pet-owners.

Many of the real or apparent differences between owners and non-owners are probably attributable to something as simple and fundamental as upbringing. In a study published in

1981,* it was found that people who had little experience of pets during childhood were far less likely, on average, to become adult pet-owners than those who did. People also tend to remain loyal to the particular species of pet they kept as children. If they had cats they become life-long cat-lovers; if they had dogs they stay loyal to dogs, and if they had both they tend to remain fairly undiscriminating.[49] While this finding both supports and contradicts the widespread folk belief that the world is divided up into dog people and cat people and 'never the twain shall meet', it also underlines another of the dangers inherent in generalizing on the nature of pet-ownership or the people who engage in this activity. It may be the case, for example, that the psychological differences between different kinds of pet-owner are greater, on average, than the differences between owners and non-owners. In a fascinating, although somewhat bizarre study carried out in California, a group of psychologists measured differences in personality between horse-, turtle-, snake- and bird-owners. They obtained some intriguing results: male horse-owners were aggressive and dominant while females were easy-going and non-aggressive. Bird-owners were socially outgoing and expressive, snake-owners were relaxed, unconventional and novelty-seeking, and turtle-owners – one almost hesitates to mention it – were 'hardworking, reliable and upwardly mobile'.[50]

It is impossible to draw any firm or definite conclusions from this maelstrom of curious and contradictory findings. It is, nevertheless, possible to say with some conviction that we have no good evidence that the majority of pet-owners are any different from anyone else, or that they use their pets as ersatz replacements for people. Rather there is a vague suggestion that some pet-owners, for reasons which are unclear, may have a greater desire for company and friendship and because of this use their pets to augment what they already derive from the companionship of humans. Having said this, however, it is

---

* These findings have been confirmed by the results of various studies published since 1981 (see e.g. Poresky et al., 1988; Kidd & Kidd, 1989).

also obvious that the massed ranks of ordinary, run-of-the-mill pet-owners conceal a few extremists who are prepared, or perhaps compelled, to carry their relationships with animals beyond the essentially arbitrary boundaries created by public opinion. Society will no doubt continue to deal with such cases according to the degree of suffering or harm they cause themselves, other people, or the animals they exploit. Either way, there are no reasonable grounds for regarding the mundane majority of pet-owners as potential zoophiles or fanatics any more than there is reason to treat all alcohol drinkers as embryonic dipsomaniacs. Pet-keeping undoubtedly does involve a degree of sentimentality, but this is not sufficient reason why the subject should be ignored or ridiculed. Much, if not most, of what people derive from close relationships involves sentiment, and there does not appear to be any obvious justification for repressing or belittling such feelings when they are applied to animals.

CHAPTER 3

Instruments of follie

I am his Highness' dog at Kew, Pray tell me, Sir, whose
dog are you?

Alexander Pope, epigram engraved on the
collar of the Prince of Wales's dog

In February 1862, during an expedition to discover the source
of the Nile, the English explorer John Hanning Speke arrived
in the Kingdom of Uganda. There he obtained an audience
with M'tesa, the twenty-five-year-old King of the large and
powerful Waganda tribe. M'tesa was a sort of nineteenth-
century Idi Amin – a despotic tyrant who delighted in
executing his subjects for trivial misdemeanours or even for
personal amusement. He was also an ardent pet-lover with a
particular fondness for dogs, especially white ones. Perhaps to
humour their murderous and capricious ruler, the other
inhabitants of the palace also kept pets in profusion. On
Speke's first visit to the King he was amazed to see courtiers
leading bulls, dogs, goats and even men and women about
attached to pieces of string, or carrying pet cockerels and
hens around in their arms.[1] The circumstances of Speke's
observations were exotic, but the phenomenon he witnessed
was not. Throughout history the world's ruling classes have
almost invariably demonstrated a powerful affinity for pets,
and this affinity has often been the excuse for mind-boggling
displays of gratuitous self-indulgence. Other examples are not
hard to find.

Nearly 800 years earlier, toward the end of the Han dynasty,
the Chinese Emperor Ling became so infatuated with his
dogs that he invested them all with the rank of senior court

officials. This entitled them to the finest food available, sump-
tuous oriental rugs to sleep on, and a personal bodyguard of
hand-picked soldiers. The Han were subsequently displaced by
a succession of foreign dynasties and Emperors who, nevertheless, followed Ling's example by displaying an inordinate
fondness for dogs, especially the pug-faced ancestors of the
modern Pekingese. The only important exceptions were
the Ming who banished dogs from court and replaced them
with cats, greatly to the annoyance of the palace dog-breeders
who until then had been doing a roaring trade supplying pets
to rich officials. When the Ming were eventually ousted by the
Manchurian Ch'ing dynasty in the seventeenth century, the
Pekingese dog was firmly reinstated in the Forbidden City
and, for the next 200 years, enjoyed a privileged status
unrivalled by any other variety of pet before or since. At the
peak of their popularity under the Manchus, the little palace
dogs were treated, quite literally, as princes and princesses,
and they received personal stipends appropriate to their
elevated station. As puppies they were suckled at the breasts
of imperial wet-nurses and as adults they were constantly
attended by a retinue of servants. To supervise their care and
husbandry the emperors created a special élite corps of royal
eunuchs who, in their heyday, exercised considerable political
influence and power.[2]

This lengthy aristocratic obsession with dogs seems to have
had little or no effect on attitudes to this species among the
ordinary working population of China. As early as 800 BC, in
the so-called *Book of Rites*, dogs were divided into three classes:
hunting dogs, watch-dogs and those known as the edible
variety. No mention was made of pet dogs.[3] Since then little
has changed except for the virtual disappearance of recognizable hunting breeds. According to modern accounts, the
average Chinese regards the dog either as a healthy and
nutritious addition to the family diet or 'purely as a burglar
alarm' which is no more a pet than 'the wretched cur chained
to a kennel in the yard'.[4] This apparent antipathy for pets
among the Chinese proletariat was more recently reaffirmed
during the years of the Cultural Revolution (1966–76) when

pet-keeping was banned as 'bourgeois', and again in 1983 when the municipal authorities in Beijing ordered the extermination or banishment of almost the entire canine population of the city – some 400,000 animals. According to reports in the British press, only police dogs, dogs used in laboratory research, and those kept under licence for eating were exempt from these draconian measures.[5]

So, for a sizeable chunk of Chinese history there has been a definite class distinction between those who did and those who did not keep animals as pets: whereas the ruling élite took their affection for dogs or cats to bizarre and outrageous extremes, the mere idea of being fond of a dog or indeed any other kind of domestic animal was a concept alien to the mentality of the majority peasant population.* Not surprisingly, given its long-standing links with Imperialism, pet-keeping acquired a distinctly tarnished image in post-revolutionary China.† Magazine articles commemorating the Chinese 'Year of the Dog' in 1982 denounced the western practice of keeping dogs for companionship as the 'product of a sick capitalist society'.[6]

China's neighbour, Japan, also had its fair share of pet-loving monarchs. The seventeenth-century Shogun Tsunayoshi, also known as the Dog Shogun, was so obsessed with canine companions that he passed a law that all dogs must be treated kindly and only spoken to in the politest of terms. Practising

---

* The only exception to this rule are pet birds. Captive songbirds have always been popular among the poorer classes of China, much to the detriment of this region's wild bird populations.

† Subsequent to the 1983 dog purges in Beijing, only privileged officials were issued with special passes that allowed them to keep dogs in the city. Despite this, illegal dog ownership has become increasingly rampant among the Chinese *nouveaux riches* during the last twelve years, along with pet-keeping in general. According to recent press reports, upwardly mobile Beijing residents will now pay as much $4,500 US for a pedigree lap-dog and up to $3,000 for a well-bred cat. Municipal dog-extermination squads still prowl the city, however, and, in a recent week-long, anti-dog drive, 351 illegal pets were beaten to death in front of their distressed owners. The official justification for these continued purges is rabies prevention. This fails to explain why dog-keeping remains legal in some cities, such as Shanghai, why cats are still legal in Beijing, or why it is considered necessary to kill the animals with such evident brutality (*Economist*, 29 Jan. 1994; Kynge, 1986).

what he preached, Tsunayoshi eventually became the proud owner of 100,000 dogs. The expense of caring for these pets over-burdened the national exchequer, caused inflation, and resulted in an unpopular new tax on farmers.[7]

Closer to home, in Europe, a similar though perhaps less extreme pet-owning class distinction has also been apparent since ancient times. The early Greek inhabitants of Sybaris, whose name has since become a byword for luxury and opulence, were ardent pet-lovers. Their favourite pets were long-haired Maltese lap-dogs, although they also enjoyed the companionship of monkeys and, intriguingly, dwarves. Maltese dogs were carried everywhere, even to the public baths and wrestling schools, and they were also encouraged to share their owners' beds at night. One of the fictional characters invented by the Greek philosopher Theophrastus (322–287 BC) kept a variety of pets including monkeys and apes, a tame jackdaw and a Maltese dog. He also purchased various toys for his jackdaw to play with and, when his dog died, buried it beneath an inscribed tombstone. The writer Lucian (AD 115–200) satirized the Greek affection for lap-dogs. In one of his comic *Dialogues* he describes a Stoic philosopher named Thesmopolis, whose mistress implores him to take care of her little pet Myrrine, because she does not trust the servants to look after the animal properly. The unfortunate gentleman is then obliged to make a complete fool of himself by carrying this dog around while it peers from the folds of his cloak, barks at passers-by, and incessantly licks his long beard. At times pets provided the Athenian aristocracy with an excuse for outrageous displays of extravagance. According to Plutarch, Alcibiades once paid seventy minæ for a dog – more than twenty times the value of a human slave – whose long and beautiful tail he proceeded to cut off simply to shock people.[8]

From roughly the third century BC onwards, birds, monkeys, Maltese dogs and even fish were common household pets among the Roman upper classes. As with most things, the Romans took their affection for these animals to even more bizarre extremes than the Greeks. Both Ovid and Catullus wrote poems to commemorate the deaths of their mistress's pet birds,[9] and the Emperor Hadrian had monumental tomb-stones erected over the graves of his favourite dogs. At one time pet turbot were all the rage in Rome. The daughter of Drusus adorned one with gold rings, while the orator Hortensius actually wept when his favourite flatfish expired.[10] Lap-dogs seem to have aroused almost slavish adulation. The poet Martial (AD 40–104) was so inspired by one belonging to his friend Publius that he waxed exceedingly lyrical on the subject:

Issa is purer than the kiss of a dove. Issa is more tender than all the young maidens. Issa is more precious than the sapphires of India. The little Issa is Publius' heart's delight; when she gives a tiny whine you would have thought that she spoke, and she knows all her master's sorrows and his joys. She lies upon his neck, and sleeps without even a sigh escaping her, and if she finds herself in need there is no fear that she would ever sully the counterpane for by a little flutter of her paw she shows she wants to be set down, and after that she asks no more than to be gently cleaned. Such is the delicacy of the chaste little dog that she loves not at all, nor is there a spouse to be found who is worthy of a nymph so tender. So that death shall never quite take her from him, Publius has had a picture painted of her just as she really is. And when you look from Issa to the painting you think you see two Issas, or two portraits.[11]

Patterns of pet-ownership in Britain followed approximately similar lines. More or less until the eighteenth century the ordinary working people of Britain were encouraged by religious and secular authorities to regard domestic animals simply as useful objects placed on earth for the economic benefit of mankind. Since their sole *raison d'être* was to serve humanity they could be used, abused or slaughtered with impunity. Such a philosophy was incompatible with the view of animals as faithful and loyal companions.[12] Yet, somewhat ironically, those in power, whether priesthood or nobility, were not always inclined to practise what they preached. When Thomas à Becket toured France in the twelfth century, as Henry II's ambassador, his entourage included a multitude of horses and hounds as well as a veritable menagerie of tame monkeys. The horses and hounds reflected Becket's obsessive interest in hunting, but the monkeys were undoubtedly pets.

Becket was by no means unusual in enjoying the companionship of monkeys. One chronicler of the period, alluding to a bishop of Durham's practice of keeping 'apes' described it as being 'the custom of modern prelates for occasionally dispelling their anxieties'.[13]

Although pet-keeping was an accepted facet of the religious life for some senior clergymen, others were evidently disgusted by the practice. Salimbene, one of the early disciples of St Francis, described it as 'a foul blemish' that many members of his order liked to 'play with a cat or a whelp or some small fowl'. Such distaste became official Franciscan policy at the General Chapter of Narbonne in 1260 when it was ruled that 'no animal be kept, for any brother or any convent, whether by the Order, or any person in the Order's name, except cats and certain birds for the removal of unclean things'.[14] Soon other religious orders began to follow suit. In thirteenth-century France, Eude Rigaud, the Archbishop of Rouen, ordered the immediate removal of pet dogs, squirrels and birds from a Benedictine convent near Evreux. Meanwhile, in Britain in 1345, Hugo de Seton, archdeacon of Ely, sent a stern letter to the abbess of another convent: 'We forbid, therefore, dogs or birds, both great and small, being kept by an abbess or any nun within the walls of the nunnery or beneath the chair, especially during divine service.'[15] In 1387 the nuns of Romsey Abbey in Hampshire provoked an even sterner rebuke from William of Wykeham for precisely the same reason:

Whereas we have convinced ourselves by clear proofs that some of the nuns of your house bring with them to church birds, rabbits, hounds and such like frivolous creatures, to which they give more heed than to the offices of the church, with frequent hindrance to their own psalmody and to that of their fellow nuns, and to the grievous peril of their souls, therefore we strictly forbid you, jointly and singly, in virtue of the obedience due to us, that from henceforth you do not presume to bring to church any birds, hounds, rabbits or other frivolous creatures that are harmful to good discipline . . . What is more, because through hunting hounds, and other dogs living within the confines of your nunnery, the alms which should be given to the poor are devoured.[16]

But even when the nuns themselves stopped keeping pets, the various ladies of noble birth who lived within their convents persisted in doing so. 'Lady Audley who boards here', wrote the unhappy sisters of one convent, 'has a great abundance of dogs, insomuch that whenever she comes to church, there follow her twelve dogs, who make a great uproar in the church, hindering the nuns in their psalmody, and terrifying them.'[17]

Naturally, the British aristocracy did not restrict their pet-keeping activities to the cloistered confines of religious seminaries. During the Middle Ages, lap-dogs, too small to have been of any value in hunting, were a popular addition to most baronial households. Noble ladies frequently carried them about in their arms, and fed them with morsels of food from the table. This habit was deplored by writers on etiquette who vainly insisted that it was impolite to fondle dogs or cats during meals. Around this time the male nobility seem to have considered lap-dogs unworthy of their attentions and, instead, lavished their affections on the more masculine and ostensibly 'useful' hunting hounds and falcons.[18] One notable exception to this was Henry III, an effete and unpopular monarch, who disliked hunting but was so attached to his favourite dog that he habitually carried it around with him in a little basket.[19]

By the sixteenth century, the fashion for lap-dogs achieved unprecedented levels of popularity, despite vitriolic criticism from William Harrison in Holinshed's famous *Chronicles of England, Scotland and Ireland*:

They are little and prettie, proper and fine, and sought out far and neere to satisfie the nice delicacie of daintie dames, and wanton womens willes; instruments of follie to plaie and dallie withall, in trifling away the treasure of time, to withdraw their minds from more commendable exercises, and to content their corrupt concupiscences with vain disport, a sillie poore shift to shun their irkesome idleness. These Sybariticall puppies, the smaller they be the better they are accepted, the more pleasure also they provoke, as meet plaiefellowes for minsing mistresses to beare in their bosoms, to keep companie withall in their chambers, to succour with sleepe in bed, and nourish with meat at bord, to lie in their laps, and licke their lips as they lie in their wagons and coches.

In the same passage, Harrison went on to deliver a scathing attack on 'people who delight more in their dogs that are deprived of all possibilitie of reason, than they do in children that are capable of wisedome and judgement. Yea, they oft feed them of the best, where the pore man's child at their dores can hardlie come by the worst.'[20]

Some high-ranking ladies of the period were clearly moved by moral diatribes of this nature. In his book *Man and the Natural World*, Keith Thomas quotes the story of Katherine Stubbes, a pious Elizabethan lady, who, on her deathbed, caught sight of her favourite puppy and promptly

beat her away, and calling her husband to her, said 'Good husband, you and I have offended God grievously in receiving many a time this bitch into our bed; we would have been loathe to have received a Christian soul . . . into our bed, and to have nourished him in our bosoms, and to have fed him at our table, as we have done this filthy cur many times. The Lord give us grace to repent it.'[21]

The majority, however, preferred to emulate sixteenth-century trend-setters such as Mary Queen of Scots, who was constantly surrounded by an entourage of tiny dogs, some of which she dressed in blue velvet suits to keep them warm in winter.[22] Mary Stuart also founded an entire dynasty of dog-loving monarchs to rival that of the Chinese Manchus. James I, Charles I, Charles II and James II and their little sister Mary were all enthusiastic dog-owners. Indeed, Charles II's fondness for dogs, particularly the little spaniels that now bear his name, was as notorious as his exploits with the ladies. Dogs overran the palace during Charles's reign, causing one courtier to remark, 'God save your Majesty, but god damn your dogs.'[23]

Perhaps the Stuarts' single-minded devotion to their dogs contributed in some way to the gradual spread of pet-keeping to the lower rungs of English society. Attitudes to animals in general, and to pets in particular, were undoubtedly changing at this time, and this change was accompanied by a progressive increase in pet-ownership among the newly emergent, urban middle-classes.[24] This national change in attitudes was charmingly reflected in the diaries of Samuel Pepys. On 12 February 1660 he wrote the following entry on the subject of his wife's dog, Fancy:

So to bed, where my wife and I had some high words upon my telling her that I would fling the dog which her brother gave her out of the window if he pissed the house any more.

Three years later on 8 April his attitude had clearly softened:

After dinner by water towards Woolwich: and on our way I bethought myself that we had left our poor little dog, that had followed us out-a-doors, at the waterside and God knows whether he be not lost; which did not only strike my wife into a great passion, but I must confess, myself also, more than was becoming me.

By 1668 his heart was won over completely:

This day, my father's letters tell me of the death of poor Fancy in the country, which troubles me, as being one of my oldest acquaintances and servants.[25]

Nowadays, of course, pet-keeping in Britain has more or less achieved full emancipation; but this does not prevent the mass media from drawing attention to the pet-loving exploits of Britain's fading aristocracy. We need only observe the unending television broadcasts depicting the present British Monarch with her ever-present coterie of Corgis.

The uppercrust and the nobility probably indulged in orgies of pet-keeping for a variety of reasons, some of them fairly objectionable. Since time immemorial people of wealth and rank have employed the services of strange and exotic creatures – both human and animal – as emblems of their lofty status; as a means of impressing others and advertising the extent of their power and affluence. Rare and unusual pets served this purpose admirably, and they had the added advantage that they could be given away as political favours or as tokens of royal esteem.[26] To a lesser degree, animals are still used for this purpose today. The well-groomed and elegant Afghan hound can fulfil the same roll as a mink coat or a Rolls-Royce; an outward expression of its owner's status and prestige.

Another equally unsavoury motive for keeping pets has been discussed in some detail by Yi-Fu Tuan in his book *Dominance and Affection: The Making of Pets*. Tuan argues that pet-keeping is basically an exercise in playful domination; a practice which stems from man's inherent insecurity, and his need to display his ability to control and subdue the unruly forces of nature. The thesis encompasses not only our treatment of animal pets, but also of ornamental gardens, topiary, bonsai trees, slaves, dwarves, fools, entertainers and even, in certain situations, women and children. The relationship, as Tuan sees it, between owner and pet is essentially patronizing, condescending, and flippant; flippant because, although pets are regarded with affection, they are also made to suffer indignities and humiliation for our entertainment, and are often casually disposed of when they cease to amuse. In support of his theory he provides an exhaustive catalogue of man's playful inhumanity to his human and non-human fellows, with particular emphasis on the genial excesses of the nobility and aristocracy.[27]

The difficulty with Tuan's idea seems to be largely semantic; it all depends on one's definition of the word 'pet'. Over the centuries, human beings, especially wealthy and powerful human beings, have undoubtedly indulged in a surfeit of the kind of *roi s'amuse* that he describes. But it is still inaccurate to argue that the thrill of dominating others is necessarily the most important thing that humans, rich or poor, derive from pet-ownership. True, pets are dependent on their owners for care and protection and this creates a fundamental inequality in the relationship. But inequality does not always entail domination. Children, after all, are dependent on their parents, but the rewards of parenthood are surely more to do with the responsibility of caring for and nurturing another individual than with domination. Animals have certainly served mankind as animated, expendable playthings. But they have also served as companions and friends. By lumping the whole lot together under one heading, Tuan has, so to speak, thrown the baby out with the bath water.

To be fair, even the ruling classes did not always keep pets solely to flaunt their power. Tuan himself, quoting Thomas Carlyle, mentions the genuinely tender feelings that Frederick the Great had for his dogs. In 1774, 'wrapped in solitude, the King shut himself up in Sans Souci with his dogs, and afterwards he asked to be buried under the terrace of this little summer palace at Potsdam among his dogs'.[28] Mary Queen of Scots provides another example. The gruesome eye-witness account of her execution describes how 'one of the executioners, pulling off her garters, espied her little dogge which was crept under her clothes, which could not be gotten forth but by force, yet afterwards would not departe from the dead corpse, but came and lay between her head and her shoulders, which being imbrued with her bloode, was carried away and washed'.[29] Going to the scaffold secretly accompanied by your favourite pet was doubtless an eccentric act by the standards of the period, but it could hardly be construed as either playful or ostentatious. Charles II, Mary's great-grandson, was prepared to undergo public humiliation for the sake of his dogs. When one of his pets was stolen he was forced to insert a distinctly peevish advertisement in a contemporary newspaper to plead for its return:*

Will they never leave off robbing his Majesty? Must he not keep a dog? This dog's place, though better than some imagine, is the only place which nobody offers to beg.[30]

It is not beyond the realms of possibility that Charles, Mary and Frederick the Great actually derived certain consolations from the companionship of animals which were lacking in their relationships with other people. In 1737, for example, the Duchess of Windsor wrote to her daughters, 'I am very fond of my three dogs, they have all of them gratitude, wit and good

---

* Intriguingly, Charles II's descendant and namesake, the present heir to the British throne, recently repeated history by placing an advertisement in the small ads section of the rather staid *Aberdeen Press & Journal* asking for infomation leading to the return of his lost dog, a Jack Russell terrier named Pooh. Needless to say, the popular press went to town on this, with some of the tabloid newspapers creating emergency 'Hot Dog' telephone lines, and offering up to £1,000 in reward. Pooh, alas, was never found (*London Evening Standard*, 21 April 1994).

sense; things very rare to be found in this country.'[31] Horace
Walpole, the Fourth Earl of Orford, made a similar point
rather more eloquently when he wrote that:

Sense and fidelity are wonderful recommendations; and when one
meets with them, and can be confident that one is not imposed upon,
I cannot think that two additional legs are any drawback. At least I
know that I have had friends who would never have vexed or
betrayed me, if they had walked on all fours.[32]

The nineteenth-century novelist, Ouida, may have hit the
nail on the head when he said that 'those who are great or
eminent in any way find the world full of parasites, toadies,
liars, fawners, hypocrites: the incorruptible candor, loyalty
and honor of the dog are to such like water in a barren place
to a thirsty traveller'.[33]

But whatever motives the ruling classes or the ecclesiastical
élite may have had for lavishing their attentions on animals,
it is clear that this long-standing association between
pet-keeping and wealth has done little to encourage a positive
view of the phenomenon. The recent proliferation of pets in
modern industrial societies is regarded by many as the direct
product of western material affluence,[34] and some would
argue that the entire practice is a self-indulgent waste of
emotional and financial resources that would be better
spent in the service of under-privileged human beings.[35]
Unfortunately, incidents that tend to reinforce such negative
perceptions still occur from time to time. Take, for instance,
the following story reported in 1983 in a popular British
newspaper:

MILADY BUYS £8,000 JET TRIP FOR DOG

Millionairess Lady Beaverbrook booked the entire business section
of a Tristar Jumbo Jet yesterday so that her dog could travel with
her . . . She said: 'No one objects when I spend 250,000 guineas on a
yearling racehorse, so why should eyebrows be raised when I spend
my money to travel in comfort?'[36]

The answer, of course, to Her Ladyship's rhetorical question
is quite simple. A yearling racehorse represents a financial

investment. Buying one, therefore, makes reasonable sense if one can afford it. But spending £8,000 for the pleasure of travelling with one's dog makes no economic sense whatsoever. It is an example of pure extravagance and one which is hardly endearing in a world where roughly one-third of the human population is permanently malnourished.*

Throughout history, aristocrats and senior clergymen, often with more money than sense, have poured sympathy and affection on their pets while, at the same time, displaying an utterly callous disregard for the unenviable plight of the working population. Their pets lived off the fat of the land while the common people often endured semi-slavery; terrorized by both church and state, and crippled by ignorance, famine and disease. Is it any wonder, then, that the grossly pampered pets of the nobility have become one of the more potent symbols of man's inhumanity to man? Or that many people still view pet-keeping as a pointless and unnecessary luxury – a frivolous invention of the idle rich – which has no place in a world where millions die each year from starvation and inadequate medical care?

The issue is an important one, but it is one which deserves careful thought. First, most of what goes into pet food is considered unsuitable for human consumption. In fact, we compete less with our pets for food than we do with, say, factory-farmed pigs that are fed largely on grain.[37] Second, the idea of discouraging pet-ownership because it is supposedly wasteful, raises a whole host of thorny moral issues. In western Europe and North America people are fortunate in living under democratic political systems which entitle them to spend their money, within reason, however they choose. Such systems allow people to engage in any number of activities which they enjoy but which are of no benefit to the mass of humanity. The billions spent each year on pet food may seem

---

* It would be nice to imagine that this kind of pet-directed extravagance had declined in recent years along with the aristocracy who traditionally perpetrated it. Unfortunately this does not appear to have been the case. In 1991, Countess Carlotta Liebenstein, an eccentric German noblewoman, left an estate valued at $80 million to a German shepherd dog called Gunther (*Reuters*, 22 January 1994).

like a terrible waste, but it is certainly no worse than many other examples of western profligacy. Should people, for instance, be entitled to drive around in expensive cars? Should they be allowed to squander money on luxury foods or fashionable clothes, or to gamble away thousands of dollars on a galloping horse or the spin of a roulette wheel? Compared with conspicuous wastefulness of these kinds, pet-keeping is positively frugal. If we are going to restrict the rights of people to own pets, then we are also morally obliged to suppress all these other activities and, in no time at all, tyranny replaces democracy.

Undoubtedly, the picture of the villainous ruler doting over some fat little lap-dog while the populace starves is a vivid and horrifying one. But there is a grave danger of allowing such potent images to distort our judgement of the whole issue. The assumption that pet-keeping is a trivial and wasteful by-product of material wealth rests on the notion that poor or non-affluent people do not keep pets. This notion is, in fact, largely erroneous. While it is generally true that wealthy individuals and societies keep larger numbers of pets and show greater affection for animals than poor ones, a great many exceptions can be found to this rule, and some of these exceptions are highly instructive. Take, for instance, the strange case of the witch's familiar.

From the middle of the sixteenth to the end of the seventeenth centuries, England became the scene of one of the more barbaric episodes in its history. During this period, a large number of innocent and often elderly people were imprisoned and condemned to death for the crime of witchcraft. Nobody really knows for certain what sparked off the witch-hunts, although most historians accept that the idea of prosecuting suspected witches was imported from continental Europe where witchcraft had long been treated as a form of heresy by the medieval Inquisition.[38]

Whoever or whatever the cause of this unpleasant business, a series of new acts were passed in England which imposed severe penalties on anyone indulging in 'necromancy', a crime which included consulting, feeding or rewarding 'any evil and wicked Spirit'. This clause introduced the concept of the witch's 'familiar' to English law for the first time, although the basic idea had already existed in popular folklore for many centuries. Familiars were regarded as supernatural companions who carried out the witch's evil intentions in return for protection and nourishment. Very occasionally, they manifested themselves as odd-looking men or children, or as creatures of monstrous or indeterminate shape, but, far more often, familiars took the form of commonplace animals such as dogs, cats, birds, mice, rats, hares, rabbits, hedgehogs, ferrets, weasels or toads. The idea of the familiar was peculiarly English; it was rarely mentioned in Continental witch-trials.[39] English witches also differed from their Continental brethren in a number of other important respects. They were almost invariably women, usually poor, and generally elderly. Often they already had a reputation for malicious or anti-social behaviour. In at least half of the well-documented witchcraft cases which were brought to trial in England, the accused was implicated by the fact that he or she possessed and displayed affection for one or more animal companions.[40] Examples are numerous, but the following transcription from a trial which took place in 1582 at St Osyth on the Essex coast is typical:

The said Thomas Rabbet saith, that his mother Ursula Kemp alias Grey hath four several spirits, the one called Tiffin, the other Titty, the third Piggin, and the fourth Jack, and being asked of what colours they were, saith that Titty is like a little grey cat, Tiffin is like a white lamb, Piggin is black, like a toad, and Jack is black, like a cat. And he saith, he hath seen his mother at times to give them beer to drink and of a white loaf or cake to eat.[41]

Ursula Kemp's only crime was a malicious tongue, loose morals and a harmless friendship with two cats, a lamb and a toad. For this she was condemned and hanged. Christina Hole, the author of *Witchcraft in England*, has summarized the situation: 'In an age when fondness for small animals was nothing like as general in England as it is now, the actual possession of any beast that might be supposed to be a familiar was a clear danger to anyone suspected of witchcraft, especially if he or she were known to treat it with affection. It was not even necessary for the creature to live in the house; a dog bounding towards a suspected person in the fields, or a cat jumping through a window might be enough to confirm an already existing suspicion.'[42]

There are two main points to be gleaned from the history of the English witch-trials. First, that old and solitary people at this time frequently cared for and nourished pet animals, despite the fact that they themselves were poverty-stricken. Second, that the savage public response which this sometimes provoked had little at all to do with economic considerations. It simply arose from the belief that this kind of relationship between humans and animals was in some way perverted and therefore wrong – a belief, it seems, which still lingers to the present day. The only reason why the aristocracy and clergy were allowed to indulge in the same practice was because, most of the time, they were quite literally above suspicion; their special status elevated them beyond the reach of public censure. And even they, as we have seen, were occasionally accused of frivolity, sacrilege, wantonness, concupiscence, irksome idleness and immoral extravagance because of their pet-keeping habits.* In other words, it was not poverty which restricted pets to the upper classes but prejudice.

A situation somewhat resembling the example of the English witches is still to be found in many modern European cities. For example, as anyone who has ever visited Rome will have noticed, the city is infested with cats. For the most part emaciated, feral animals which live in loose colonies and eke out a meagre existence around the city's gardens, parks and ruins. According to an Italian zoologist who has studied these cats in some detail, most of the colonies are supported by one

---

* Under certain circumstances, pet-keeping aristocrats were also accused of witch-craft. During the English Civil War (1642–6), Puritan pamphlets declared that Prince Rupert's pet poodle, Boy, was in fact a familiar, and that this dashing young Cavalier commander used the animal's supernatural powers to render himself 'shot free' (impervious to bullets) during battle. According to one such pamphlet, the fortunes of the War finally turned in the Roundheads' favour when Boy was deliberately shot and killed with a silver bullet on the battlefield of Marston Moor in 1644 (Dale-Green, 1966: 81–3).

or more human benefactors who regularly bring the animals handouts of food. Most, but not all, of these benefactors are women, the majority are elderly and have few living friends or relatives. All of them are poor and, in some cases, destitute. Fortunately, they are not in danger of torture or execution. They are, nevertheless, regarded as insane by the majority of their countrymen.[43]

In his 1947 essay *The English People*, George Orwell professed himself baffled by the way in which poverty did nothing to curb the Englishman's obsession with pets:

Although its worst follies are committed by upper-class women, the animal cult runs right through the nation and is probably bound up with the decay of agriculture and the dwindled birthrate. Several years of stringent rationing have failed to reduce the dog and cat population, and even in poor quarters of big towns the bird fanciers' shops display canary seed at prices ranging up to 25 shillings.[44]

Orwell concluded that sentimentality about animals (together with suspicion of foreigners and an obsession with sport) was simply a curious national characteristic which singled the English out from other peoples. Doubtless, he would have been even more perplexed had he realized the full extent of the pet-keeping habit in cultures far less materially affluent than his own.

CHAPTER 4

Pets in tribal societies

Savages may be brutal, but they are not on that account
devoid of our taste for taming and caressing young
animals.

Sir Francis Galton, *Inquiry into Human Faculty*, 1883

At about the same time that the English were torturing and
murdering elderly pet-owners for the crime of necromancy
during the sixteenth century, a Spanish explorer called
Francisco Hernandez was busily recording his first
impressions of the newly discovered province of Nova Hispania
– a region we now call Mexico. Hernandez took it upon
himself to describe the strange and unfamiliar creatures he
encountered in the new territory, and one of the first to attract
his attention was a curious beast he found living in the homes
of the Indians. 'The *Mapach*', he wrote 'is an animal of
dimensions a little larger than a Maltese dog. Short, rotund
and hairy with indistinct black and white patches, a large
head, small ears; a dog-like muzzle, long and fine, and with
some white stripes which run from the eyes to the shoulders; a
long tail and human-like feet and hands with which it appears
to feel everything. Domesticated and fed in the house, it is
constantly pestering people it knows and will follow them with
great affection. It lies next to them and rolls around happily in
the soil, amusing itself and gambolling in a thousand different
ways.'[1]
   Judging from the description and accompanying illustration, Hernandez's *Mapach* was none other than the raccoon
(*Procyon lotor*), a species which was also kept as a house-pet by

the Indians of California,[2] and one which was until recently
valued for the same purpose by an estimated 1.8 million
modern Americans.[3]

The Indians of North America also made pets of animals as
large as moose. The eighteenth-century traveller, Hearne,
reported having 'repeatedly seen moose at Churchill as tame
as sheep and even more so'.

> The same Indian that brought them to the Factory had, in the year
> 1770, two others so tame that when on his passage to Prince of
> Wales's Fort in a canoe, the moose always followed him along
> the bank of the river; and at night, or on any other occasion that the
> Indians landed, the young moose generally came and fondled on
> them, as the most domestic animal would have done, and never
> offered to stray from the tents.

Similarly, Sir John Richardson observed that the Indians used
to capture and keep bison calves, wolves and other animals,
and it was not unusual for them 'to bring up young bears, the
women giving them milk from their own breasts'. On
purchasing one of these bear-cubs he wrote that 'the red races
are fond of pets and treat them kindly; and in purchasing them
there is always the unwillingness of the women and children to
overcome, rather than any dispute about price. My young bear
used to rob the women of the berries they had gathered, but
the loss was borne with good nature.'[4]

In other parts of the New World different kinds of companion animals were favoured. According to one early account,
the natives of the West Indies kept 'small cur dogs' which
'never bark nor do anything except eat and drink'. Although
these dogs may not have been much use in the practical sense,
they presumably made good pets since they are reported as
showing 'deep affection toward those who feed them, and wag
their tails and frisk about merrily, showing that they want to
please the one who feeds them and the one they regard as
master'.[5] A later visitor also mentions that the inhabitants of
St Domingo (Jamaica) were so fond of their little dogs that
they carried them on their shoulders wherever they went and
'nourished them in their bosoms'.[6]

For the Indians of South America animal taming and

pet-keeping were practically minor industries. The eighteenth-century Spanish explorers, Juan and Ulloa, were plainly astonished by the degree of affection shown for these animals:

Though the Indian women breed fowl and other domestic animals in their cottages, they never eat them: and even conceive such a fondness for them, that they will not sell them, much less kill them with their own hands. So that if a stranger who is obliged to pass the night in one of their cottages, offers ever so much for a fowl, they refuse to part with it, and he finds himself under the necessity of killing the fowl himself. At this his landlady shrieks, dissolves into tears, and wrings her hands, as if it had been an only son.[7]

The English naturalist, Bates, who travelled extensively in the Amazon region during the nineteenth century, furnished Sir Francis Galton with a list of 'twenty-two species of quadrupeds that he has found tame in the encampments of the tribes of that valley'.[8] Likewise, in a compendious review of the arts and customs of the Guiana Indians, the anthropologist Roth describes how: 'Women will often suckle young mammals just as they would their own children; e.g. dog, monkey, opposum-rat, labba, acouri, deer, and few, indeed, are the vertebrate animals which the Indians have not succeeded in taming. It is the women who especially cultivate the art of bird taming, some of them holding quite a reputation in this respect.' He then concludes by stating that: 'a native will never eat the bird or animal that he has himself tamed any more than the ordinary European will think of making a meal of his pet canary or tame rabbit'.[9]

Wholesale pet-keeping has undergone a decline in South America, along with the native people who practised it. Nevertheless, recent or contemporary examples can still be found in certain areas. During the 1930s, the English adventurer Peter Fleming observed that the Caraja people of south-eastern Brazil were 'devoted to their children and their pets'.

The villages swarmed with livestock. At nightfall parrots warred with scrawny poultry for roosts along the roof-pole. Pigs, and dismal dogs, and fantastically prolific cats, and tame wild ducks wandered in and out of the huts through holes in the wall. In almost all the northerly villages cormorants paddled among the litter round the cooking fires; sometimes their sombre plumage had been decorated by the children with tufts of red arara's feathers fastened to their wings . . . I told the Indians that they could train their cormorants to bring them fish if they fastened rings round their necks. In conception, rather than in execution, this project amused them very much; it is clear that they thought of the birds always as guests, never as servants.[10]

Fleming went on to say that the Indians 'were very fond of all these creatures, and treated them well; they asked prohibitive prices for their parrots, and the big red ones they would not sell at all'. The Warao who live around the mouth of the Orinoco River in Venezuela still keep a wide variety of species around their homes, including wild birds, monkeys, sloths, dogs, rodents, ducks and chickens. All are kept 'for sheer enjoyment, never is one eaten'.[11] The Kalapalo Indians of Brazil specialize in bird taming, and the special relationship they have with their pet birds has been the subject of study. Basso, an anthropologist who lived with the Indians, says that:

The *itolugu-oto*, 'pet-owner' relationship is characterised on the human side by nurture and protection within a household, and on the avian side by lack of *ifutisu* (in the sense of shyness). In other words, by tameness. This relationship is particularly interesting because the distinctive features are also those which define the filiative relationship, or that between human parents and their children. Children and pets alike are ideally supposed to be fed, reared and kept protected within the confines of the house. Often pets are kept secluded like human adolescents 'to make them more beautiful'.[12]

Also, according to Basso, pets are never eaten or killed, although they may belong to species which are classified as edible, and when they die they are often buried close to the house or hammock of the former owner. All of which seems to indicate that the Kalapalo affection for birds is not fundamentally different from the European's devotion to his pet dog or cat.

Among the Barasana Indians of eastern Colombia,

pet-keeping remains one of the principal leisure activities. Small rodents, dogs, parrots, and an enormous variety of other large and small birds are the most frequent pets, although tapir, peccary, ocelot, margay and domestic cats are also kept in small numbers. With the exception of the carnivores, virtually all of these animals are also hunted for food, but as pets they are never slaughtered or consumed. Again, women are the chief animal-tamers and pet-owners, but some men, particularly shamans, are equally enthusiastic. One shaman, for example, created quite a reputation for himself by taming and keeping a jaguar as a pet. In general, the Barasana look after their pets extremely well. Women will suckle puppies and hand feed other young mammals, even masticating plant foods such as manioc and banana for their tame parrots and macaws. One individual was also observed to spend hours catching small fish to feed a pet kingfisher. Occasionally, parrots and other birds have their tail-feathers plucked to provide the Indians with decorative plumes but, according to Stephen Hugh-Jones who has studied the Barasana for many years, pet-keeping is not motivated by economic or practical considerations. These people simply enjoy the business of looking after and nurturing their pets. The animals are a continual source of discussion and entertainment, and are regarded as an integral part of the community.[13]

Obvious examples of pet-keeping among the tribal peoples of the Old World are less common, but still evident. Galton wrote that the Aborigine women of Australia 'habitually feed the puppies they intend to rear from their own breasts, and show an affection for them equal, if not exceeding, that to their own infants'.[14] Similar observations were made by the nineteenth-century Swedish explorer Carl Lumholtz. He noted that the Aborigines rear dingo pups 'with greater care than they bestow on their own children. The dingo is an important member of the family; it sleeps in the huts and gets plenty to eat, not only of meat but also of fruit. Its master never strikes, but merely threatens it. He caresses it like a child, eats the fleas off it, and then kisses it on the snout.'[15] The Stone-Age Andaman Islanders displayed a similar fondness for dogs. According to Cipriani, their 'inordinate love of dogs has allowed the animals to become pests. They already considerably outnumber the human population; families of three or four people may have ten or twelve dogs.'[16] Pets are (or were) also common in the camps of the Semang Negritos of Malaysia, and young animals such as pigs and monkeys are frequently suckled by the women. One early anthropological observer reported seeing a woman with 'a child at one breast and a monkey at the other' and he also mentions that 'young animals reared in this way are not killed'.[17]

The native Polynesians were obsessively interested in pet-ownership. The Fijians apparently made pets of fruit bats, lizards and parrots, and in Samoa, pet pigeons and eels were great favourites. One early account describes how pet eels were tamed and fed until they reached a prodigious size:

Taoarii had several in different parts of the island. These pets were kept in large holes, two or three feet deep, partially filled with water. I have been several times with the young chief, when he has sat down by the side of the hole, and by giving a shrill sort of whistle, has brought out an enormous eel, which has moved about the surface of the water and eaten with confidence out of his master's hand.[18]

During a visit to Hawaii in 1773, George Forster wrote: 'the dogs in spite of their stupidity were in high favour with the women who could not have nursed them with a more ridiculous affection if they had really been ladies of fashion in Europe'.[19] Another visitor noted that 'every woman has a pet animal; and mothers who are nursing their offspring will suckle a puppy at the same time in a rivalry by no means in favour of the strength or number of their own progeny'.[20] Such devotion to animals was not evidently confined to women. According to the anthropologist Luomala, affection for selected pets was part of a marked Polynesian trait, 'men, women and children, of all social ranks, fondled, pampered and talked to their pets, named them, and grieved when death or other circumstances separated them'. She also records that grief over the death of a pet dog was revealed through 'tears and poetical eulogies' and that such animals were sometimes 'given special burial, a further indication of its owner's esteem'.[21]

The existence of pet-keeping among so-called 'primitive' peoples poses a problem for those who choose to believe that such behaviour is a pointless, modern extravagance; a mere by-product of western decadence or bourgeois sentimentality. Doubtless, when one looks exclusively at pet-keeping in prosperous cultures such as our own, it is easy to conclude that the practice is a manifestation of some eccentric cultural aberration. The fact that we squander vast resources on the habit is also of little significance since conspicuous waste is a common feature of our society. But this line of reasoning runs into serious difficulties when we contemplate precisely the same phenomenon among, say, the Semang Negritos or numerous Amerindian groups. These people are predominantly hunter–gatherers or subsistence horticulturalists and, while they may indeed have time on their hands to engage in certain leisure activities, they are not in a position to waste resources on gratuitous luxuries. Nevertheless, they seem to be prepared to invest as much time, energy and emotion in economically useless pets as the average middle-class European or North American.

Perhaps because it appears superficially anomalous, anthropologists have devoted extraordinarily little attention to the subject of pet-keeping in tribal societies. And where they have discussed it, they have tended to turn the problem on its head. Rather than tackling the reasons why such people should keep pets at all, they have concentrated, instead, on the question of why they do not kill them and eat them, as if the only sensible or understandable reason for keeping and caring for an animal is in order, ultimately, to devour it.[22] Some have argued, for example, that people avoid killing and eating pets because of the animals' resemblance to humans, and their inclusion in the social world of people. Others have pointed out the symbolic association between the act of eating a pet and the act of sexual intercourse between close relatives. We do not eat our pets, so this story goes, because it would be metaphorically equivalent to committing incest.[23] And some have proposed more general theories along the same lines. According to Edmund Leach, people feel anxious about (and therefore taboo) things which are hard to classify – e.g. things which have intermediate or ambiguous properties.[24] Pet-eating is taboo because pets straddle that uncertain, ambiguous territory between humans and animals. They are neither strictly one thing nor the other.*

Ideas like these that focus on the animal's symbolic connotations have provoked a down-to-earth, materialistic reaction from certain quarters. Marvin Harris, a 'cultural materialist', suggests that our reasons for not eating pet animals such as dogs are nothing whatever to do with this creature's symbolic or metaphorical status. They simply express the practical and economic difficulties associated with farming dogs for food. Dogs are carnivores and, therefore, grow and reproduce best on a diet of meat. Because of this, the commercial production of dogs for food would be hopelessly uneconomical. The sensible, practical solution to this problem is to do precisely what the majority of cultures do: namely farm herbivorous and omnivorous livestock such as cows, sheep and pigs which are more cost effective.[25] But the American structuralist Marshall Sahlins counters this argument by asking a similar question about the horse. After all, horses are herbivores, and they are also regularly eaten in some European countries. Yet, in Britain and the United States, horse-eating is virtually taboo. To quote from Sahlins's book *Culture and Practical Reason*:

Dogs and horses participate in American society in the capacity of subjects. They have proper personal names, and indeed we are in the habit of conversing with them as we do not talk to pigs and cattle . . . But as domestic cohabitants, dogs are closer to men than are horses, and their consumption is more unthinkable: they are one of the family. Traditionally horses stand in a more menial, working

---

* For some detailed critiques of Leach's thesis, see Halverson (1976) and Serpell (1990a).

relationship to people; if dogs are as kinsmen, horses are as servants and nonkin. Hence the consumption of horses is at least conceivable, if not general, whereas the notion of eating dogs understandably evokes some of the revulsion of the incest tabu.[26]

In Sahlins's view, the edibility of an animal is inversely related to its humanity and little, if anything, to do with its practical utility as a food item. Unfortunately, he fails to say why human beings show this peculiar tendency to befriend certain animals; to personalize them, and to treat them as members of the family.

An alternative to these intriguing speculations has been to argue that pet-keeping among tribal peoples does, after all, serve a useful purpose, albeit a somewhat trivial one. When studying such societies, several anthropologists have observed returning hunters bringing small wild animals back alive and promptly turning them over to their children. Like Christmas gifts in our society, these animated toys are generally badly treated, short-lived and, more often than not, end up the objects of target practice or mutilation. According to the anthropologist Laughlin, children receiving such 'pets' and having the opportunity to play with them are acquiring valuable educational experience both of animals and of animal behaviour – experience which may one day help them to become better hunters themselves.[27] Unfortunately, this interesting idea suffers from the same sort of problems as Yi-Fu Tuan's theory discussed in the previous chapter. It is easiest to illustrate this point with a suitable example.

Downs, an anthropologist who studied the Navaho Indians, informs us that their most usual pet is 'the house cat, which, nevertheless, is not at all common, with perhaps only one in four or five households boasting a cat. Although a cat may serve to keep down the rodent population, this service is certainly not a primary reason for keeping one.' Downs fails to elaborate on what the primary reason is, but he does say that the cat generally belongs to one particular individual, that it is allowed to sleep in the house or hogan, and that its owner will devote considerable efforts to protecting it from marauding dogs; all of which suggests that the Navaho are deeply attached to their cats. Two paragraphs later, he also observes that 'should a hunter or rambling children come upon a wild creature that can be taken alive, it will be captured and may play a brief role as a pet. Usually such captures are made by young boys who take baby rabbits, chipmunks, ground squirrels, horned toads, birds or lizards and handle them roughly as short-lived pets.'[28] Downs refers to both classes of animals – the cats and the lizards – as 'pets' when, in this particular case, they are fundamentally different. The latter constitute ephemeral playthings which may or may not serve as useful childhood instruction. The former are cherished possessions and the objects of strong, emotional attachments. By applying the blanket term 'pet' to both adored companions and expendable toys, the issue is merely confused.

Yet, when dealing with non-western cultures, gratuitous affection for animals often appears so out of place that many observers seem obliged to subject the phenomenon to practical or economic interpretation, as if no other motives could exist. The explorer, Carl Lumholtz, for instance, was clearly puzzled by the Australian Aborigines' affection for dingoes, but he managed to rationalize things by noting that this species 'is very useful to the natives, for it has a keen scent and traces every kind of game'.[29] In other words, as far as Lumholtz was concerned, the Aborigines loved their dingoes because the dingoes were useful hunting partners. Modern examples of precisely the same kind of logic are widespread. Cipriani explains the Andaman Islanders' inordinate fondness for dogs by stating that 'dogs meant invariable and abundant success in the hunt, and it is hardly surprising that the Andamese came to develop a somewhat unbalanced affection for them'.[30] Another good example is provided by a study of the Punan Dyaks of north Borneo. Here, according to the anthropologist Harrison, over a period of five days for which he kept records, one group of Punan with the aid of their dogs 'killed thirty wild pig, several of them enormous and fat'. Then, like the previous author, he states that 'in return, these Punans literally *love* their dogs. Both men and women, but particularly the latter, can be seen at any time carrying a favourite dog around, usually held loosely under one arm or across the buttocks, hands stretched back to support the dog lightly, its head sticking out to the left and tail to the right behind.'[31]

This kind of reasoning is peculiar. After all, no normal human being would give, say, a can-opener a personal name, caress it lovingly, or go into mourning when it ceased to function just because it provided a convenient method of gaining access to food. Why then should we expect tribal peoples to behave in this way toward their dogs. In reality, of course, they do not. The Aborigines inhabiting the Northern Territory and the central desert regions of Australia also keep dogs and dingoes in large numbers and give every indication of being just as fond of them as the Aborigines encountered by Lumholtz in Queensland. In spite of this, they make little use of these animals for hunting or for any other obvious purpose.[32] Conversely, the Barasana who lavish attention on economically valueless, wild animal pets, regard some of their dogs as disgusting, disreputable creatures and frequently subject them to pointless physical abuse. Yet dogs are useful trade items, and they are also of practical value as hunting partners.[33] Similarly, the B'Mbuti Pygmies, a hunter–gatherer group from the forests of Zaire, make very effective use of dogs for hunting a variety of game. Some authorities have even suggested that they would be unable to hunt properly without them. Yet they treat these dogs so aggressively and brutally that the anthropologist Schebesta was forced to remark: 'I thank God we are not pygmies. I thank him still more that we are not pygmy women, and even still more again that we are not pygmy dogs.'[34]

An even better illustration is provided by the tough and war-like Comanche Indians who lived on the Great Plains of North America. According to Linton, the economy of the Comanche was based on buffalo-hunting and raiding neighbouring tribes, and both of these activities depended on horses and horsemanship. Yet the Comanche treated their horses in much the same detached and unemotional way that we treat utilitarian objects such as bicycles. In contrast, Comanche dogs were of no practical or economic value at all and were kept simply as pets, yet Comanche warriors regarded the loss of a dog as far more devastating than the loss of several horses, and most of them would spend hours discussing with fondness all the dogs they had ever owned.[35] In other words, the Comanche were no different from us: they treasured and doted on their useless dogs, and treated their indispensable livestock like consumer goods. The ways in which people treat their domestic animals are not independent of practical considerations, but it is difficult to find any evidence at all that we love them in proportion to their usefulness.

Finally, some mention should be made of an imaginative theory proposed by the French anthropologist, Philippe Erikson, who has suggested that Amazonian hunter–gatherers keep pets primarily as a means of currying favour with wild animals and their spirit guardians. According to Erikson, the animist beliefs of Amazonian hunters render them constantly fearful of supernatural revenge or retribution for the act of killing and eating wild animals. Adopting and caring for young wild creatures, he argues, represents a way of reducing this fear by, so to speak, reimbursing the aggrieved animal spirits with kindness directed toward specific, individual pets of the same species.[36] It is unclear from Erikson's account whether he derives this charming idea from conversations with Amerindian informants or has generated it himself. Either way, it has the ring of a *post hoc* explanation. Amazon hunters may occasionally justify the pet-keeping habits of their women and children as a means of keeping the spirit world happy, but it is hard to imagine the practice evolving and persisting purely as a form of superstitious insurance policy. As Stephen Hugh-Jones's work with the Barasana indicates, pet-keeping is primarily a leisure activity. People do it because they seem to derive genuine pleasure and enjoyment from the acts of caring for, and interacting with, animal companions.*

Apart from demonstrating that it is not confined to

---

* Ironically, fear of spiritual retribution may lead some hunter–gatherer groups to precisely the opposite view of pet-ownership. Among the Koyukon of Alaska, pet-keeping is largely proscribed because the animal spirits might consider such behaviour disrespectful (Nelson, 1986: 219).

decadent western societies, the widespread occurrence of pet-keeping among tribal peoples is also of considerable historical significance. If we go back about 13,000 years to the end of the last Ice Age, before the domestication of animals or the earliest cultivation of plants, the entire human population of this planet lived more or less like the Andaman Islanders, the Semang, or the Australian Aborigines, by hunting, fishing and foraging for wild foods on a day-to-day basis. Like their more recent counterparts, it is highly likely that these Ice-Age hunter–gatherers also kept, suckled and cared for a variety of wild animal pets, including the ancestors of some of our oldest domestic species. If we now postulate that some of these pets managed to breed in captivity, and assuming our forebears overcame their scruples about eating them and exploiting them in other ways, then we are already well on the way toward true domestication.[37] In other words, prehistoric pet-keeping may well have paved the way toward animal husbandry and livestock farming.

Archaeologists even have some tentative evidence to support this theory. In 1978, at a late Palaeolithic site in northern Israel, a unique human burial was discovered. The tomb contained two skeletons: that of an elderly human of unknown sex and, next to it, the remains of a five-month-old domestic dog. The two had been buried together roughly 12,000 years ago. The most striking thing about these remains was the fact that whoever presided over the original burial had carefully arranged the dead person's left hand so that it rested, in a timeless and eloquent gesture of attachment, on the puppy's shoulder.[38] The contents of this tomb not only provide us with some of the earliest solid evidence of animal domestication, they also strongly imply that man's primordial relationship with this particular species was an affectionate one. In other words, prehistoric man may have loved his dogs and his other domestic animals as pets long before he made use of them for any other purpose. Affection for pets may seem, in retrospect, trivial and unimportant. Yet it may have been responsible for one of the most profound and significant events in the history of our species.

CHAPTER 5

A cuckoo in the nest

When you like children and dogs too much, you love them
instead of adults.

Jean-Paul Sartre, *Words*

Atemeles pubicollis is a small, rather nondescript beetle which is
found throughout much of Europe. Although its appearance
is unremarkable, *Atemeles* is an accomplished social parasite
which has perfected the art of living as a permanent guest
inside ant colonies.

Being invited to stay in an ant's nest is no easy matter. Ants,
understandably, are not very friendly toward intruders and
generally guard their nests well. *Atemeles* however has overcome this problem by evolving an intricate set of what one
might call 'behavioural keys' which help it to unlock the ants'
defences. Once the beetle has located a suitable ant colony,
which it does by scent, it wanders around outside the entrance
until it encounters a passing worker ant. As soon as this
occurs, the beetle quickly turns around and presents the ant
with a gland situated at the tip of its abdomen which secretes
a sort of ant 'narcotic'. The ant consumes some of the
substance and is apparently subdued by it. Next, a second
gland on the side of the abdomen switches on and begins
secreting a chemical which seems to operate like the insect
equivalent of a letter of introduction. No sooner has the
worker sampled it than she immediately picks the *Atemeles*
gently up in her mandibles and carries it deep inside the nest.
Even in the unlikely event of the ant attacking, all is not
lost. *Atemeles* also possesses a secret weapon in the shape of a

third abdominal gland which instantly exudes a powerful ant repellent.

Having infiltrated a colony, the beetle's troubles are more or less over. Should it feel hungry, it has only to tap a passing worker with its legs or antennae for the latter to stop and regurgitate food like an animated slot-machine. It even gets away with murder. The larvae of *Atemeles* secrete a chemical which, to the ants, smells exactly like the colony's own brood. A worker encountering one of these larvae will immediately pick it up and 'return' it to the nursery where it will promptly set to and devour the colony's own eggs and larvae. *Atemeles* contributes precisely nothing to the well-being of the ants. In fact, it is a nuisance and a burden on the colony. Yet the workers not only tolerate the beetle and its larvae, they actively encourage them to live in their midst.[1]

The entomologist Wheeler once remarked that if people were to behave like the victims of *Atemeles*, 'we should delight in keeping porcupines, alligators, lobsters, etc., in our homes, insist on their sitting down to table with us and feed them so assiduously with spoon victuals that our children would either perish of neglect or grow up as hopeless rachitics'.[2] Curiously, he did not at the same time point out that this is really only an exaggerated description of the way in which many people *do* behave with their pets. True, you will not find many alligators or lobsters curled up on suburban hearthrugs. Nevertheless, studies have shown that the vast majority of western pet-owners regard their pets as members of the family; that they talk to them, share their meals with them, allow them to sleep on the bed, and to sit on the furniture, and even celebrate their birthdays.[3] And, very occasionally, people do appear to display genuinely greater affection for their animals than for their own offspring. Surely, if all this is true, then a case exists for arguing that pet-owners are not substantially different from ants; that they are unwittingly playing host to an artful collection of social parasites.

Of course, people who keep pets may understandably object to being compared with insects. After all, ants are not exactly noted for their intelligence. Rather they seem to operate like robots; programmed to respond in an inflexible, all-or-none fashion to a limited set of chemical and tactile signals. This lack of flexibility makes them inherently vulnerable to the kind of simple deception perpetrated by *Atemeles*. Such a thing, one would imagine, could never happen to a human society. Humans are not just mindless automata like ants. We are intelligent, thinking beings, far too clever ever to be taken in by such insidious trickery. Or are we? Although humans are intelligent and adaptable in ways which ants are not, and although most of what we do is determined by learning and experience rather than blind instinct, certain areas of our behaviour are probably influenced by biological predispositions. One area in particular which has received a lot of attention from scientists concerns our responses to infants and young children.

Human infants look and behave quite differently from adults or even older children, and there is a strong body of scientific opinion which argues that these peculiarities of appearance and behaviour have evolved specifically to elicit parental care and protection. The manner in which the majority of adults respond to infants seems to bear out this idea. People do, in general, find babies appealing and attractive, and they show a strong tendency to smile at, touch, caress, embrace and care for them in highly stereotyped and consistent ways.[4] It is even possible to provoke stronger than normal responses to infants by artificially accentuating their babyish qualities. Psychologists, for example, have shown adult human subjects illustrations of both 'normal' and 'super' babies in which the infantile facial characteristics have been artificially exaggerated. When asked to choose the picture that looked most like a real baby, most of the subjects showed a preference for the super baby over the normally proportioned, real one.[5]

This attraction to baby-like creatures appears to be so strong that it is even released by the young of other species. Place a puppy or a kitten on someone's lap and nine times out of ten you will provoke a stock emotional reaction: the person will tend to caress and fondle the animal, bring it close to his or her face, look into its eyes, cuddle it, and accompany all these stereotyped actions with a chorus of incoherent, but equally stereotyped, verbal endearments. If the animal yelps or mews or shows other signs of distress, the person will display concern and do his or her utmost to comfort it. In other words, just the sight of the baby animal seems to be enough to override normal behaviour and replace it with something closely resembling the sorts of actions and activities that human parents generally direct toward their own infants. Biologists have coined the phrase 'cute response' to describe this reaction.[6]

The Nobel Prize-winning ethologist Konrad Lorenz has explained the cute response in the following terms: when one compares the majority of baby vertebrate animals – be they birds, cats, dogs, humans or even Komodo dragons – one finds that they all share certain physical features in common. Their heads are far larger in proportion to their bodies than adults, their limbs are shorter and chubbier, their eyes are bigger, and their jaws, mouths and noses are smaller and less protuberant. Overall, they have a fatter, rounder appearance; softer skin, scales, fur or feathers, and clumsier, less-coordinated movements. Animals or representations of animals displaying some or all of these characteristics tend to be perceived as 'cute' and, according to Lorenz, are able to release an *innate* desire to protect and nurture. This response has evolved, so he maintains, to ensure that human parents respond appropriately to the sight and sound of their own infants.[7] The fact that it is also released by less appropriate objects is merely a trivial flaw in a system which is otherwise efficient and highly adaptive. The Harvard biologist, Stephen Gould, puts the case rather more bluntly when he states that we are 'fooled by an evolved response to our own babies, and transfer our reactions to the same set of features in other animals'.[8] He also suggests that Walt Disney cartoonists and doll and toy manufacturers have been aware of this for some time and now design their products to create maximum appeal by exaggerating their infantile features.

Humans are not the only animals to respond parentally to

the infants of other species. Many animal parents can be persuaded to adopt and foster alien youngsters and, in nature, this lack of discrimination has led to some bizarre parasitic relationships. The European cuckoo, for example, is a 'brood parasite'. It specializes in persuading other species – chiefly small songbirds such as warblers – to rear its offspring. The cuckoo achieves this deception, first, by disguising her eggs to look like those of the host species and, second, by producing a youngster which, having evicted its foster-siblings, then mimics and exaggerates the features which parent birds find most appealing about their own progeny. The most important of these features, from the cuckoo's viewpoint, is the bright pink lining of its enormous, gaping beak which resembles a larger-than-life replica of a normal nestling's food-begging gape. As if bewitched by this alluring spectacle, the unfortunate parent birds redouble their foraging efforts in order to keep the parasite well fed and contented.[9] The host species derives no benefit at all from this relationship. Like the ants, the birds are simply duped by their own inflexible parental instincts into caring for anything which imitates or enlarges upon the signals ordinarily transmitted by their own young.

The human cute response – this habit people have of caring for small, cuddly animals – bears more than a superficial resemblance to the songbird's infatuation with its parasitic cuckoo nestling. And this has led some experts to argue that the entire phenomenon of pet-keeping is nothing more nor less than an elaborate case of social parasitism. Needless to say, this idea has done little to promote a positive view of pets or their owners. Rather, it creates the impression that pet-owners are the victims of some kind of bizarre affliction, and that dogs, cats and budgerigars are little different from body lice, fleas or tapeworms or, indeed, any other sort of parasitic organism.

Whether or not one agrees with this point of view, there is undoubtedly abundant circumstantial evidence that people tend to perceive and treat their pets as if they were children. This tendency was explored in some detail by a study at the University of Leicester in England using a standard method of psychological assessment known as Role Construct Repertory Grids or 'Rep. Grids' for short. Briefly, the Rep. Grid technique provides a method of revealing how people unconsciously perceive, construe and interpret the world in which they live, and it is widely used in a variety of clinical and occupational contexts. In the Leicester study, thirty subjects were tested with Rep. Grids in order to compare their perceptions of the owner–pet relationship with their perceptions of relationships with other people. The other people chosen for comparison were, respectively: a parent of the same sex as the subject, the subject's spouse or girlfriend/boyfriend, a friend of the same sex, a child (less than ten) known personally to the subject, the subject's own child (when less than ten), and someone the subject does not get on with. To cut a long story short, the results showed clearly that the only relationship which was seen as similar to that with a pet was the relationship with a child, principally one's own child. It was also clear that people tend to see pets as child-like, regardless of their own age or sex.[10]

In many ways this result is hardly surprising considering the number of different components the owner–pet and parent–child relationships share in common. In their book *Between Pets and People*, Alan Beck and Aaron Katcher list some of the more obvious similarities:

Like children, the animal must be continually cared for: fed and watered, kept from eating dangerous foods and objects, bathed, groomed, protected against the elements, clothed when necessary, brought to the doctor and spoken for at the doctor's. Like children, pets are petted, stroked and touched at the will of the owner. The pets range of motion is curtailed to protect it from harm, and its sexual expression is controlled and limited. However, the act that critically defines a pet as a child is our willingness to put up with the excrement of cats and dogs: to handle it, permit it in the house, to accept it in the streets.[11]

People even talk to pets and children in similar ways. Psychologists have found that, in terms of its structure, the manner in which people talk to their dogs strikingly resembles the specific kind of language, known as 'Motherese', which is adopted by mothers when talking to their infants.[12] Pets also possess the added attraction that they never, in a sense, grow up. Real children develop, become independent, and leave the nest. But the pet, throughout its life, remains in a perpetual state of innocent, infantile dependence; a fact which led Aaron Katcher to refer to them as 'icons of constancy' in an ever-changing world.[13]

Other less direct, but nevertheless suggestive, evidence also points to the potential child-like connotations of pet animals in western society. In a study of cat-owners, childless owners tended to regard their pet as a child more frequently than owners with children.[14] Similarly, in a comparable study of dog-owners, the tendency to see the pet's role as equivalent to that of a child was commoner among young and old childless couples, as well as among widowed, divorced and separated individuals.[15] Pet-owning couples without children also interact more with their pets, on average, than those with children, and also seem to become more attached to them.[16] All of which lends some credence to the popular belief that pets occasionally serve as 'child substitutes' for people who, for whatever reason, have no other outlet for their maternal or paternal inclinations. Anecdotally, of course, many pet-owners will candidly admit to using pets for this purpose and, according to one London veterinary surgeon, 'one of the most common problems seen by vets is the over-indulged pet. The owners are frequently either childless or empty-nesters, and the pet is treated as a spoiled child.'[17]

Judging from the few case histories that find their way into the psychiatric literature, this habit of using pets as surrogate children can, in rare cases, lead to pathological conditions in which patients sacrifice everything, including family, friends and self-respect, for the sake of their animals.[18] One is inevitably reminded of the English witches; elderly, mistrusted and lonely, finding solace in the care and nurturance of their pet 'familiars'. Indeed, historically, the notion of an animal serving as a child substitute is an ancient one which even features in the writings of Plutarch. In his *Life of Solon* he refers to:

the soul, having a principle of kindness in itself, and being born to love as well as perceive, think or remember, inclines and fixes upon some stranger when a man has none of his own to embrace. And alien or illegitimate objects insinuate themselves into his affections, as into some estate that lacks lawful heirs; and with affection come anxiety and care; insomuch that you may see men that use the strongest language against the marriage-bed and the fruit of it, when some servant's or concubine's child is sick or dies, almost killed with grief, and abjectly lamenting. Some have given way to shameful and desperate sorrow at the loss of a dog or horse.[19]

The same theme also appears towards the end of William Harrison's sixteenth-century diatribe against ladies of noble birth and their devotion to lap-dogs: 'the former abuse peradventure reigneth where there hath been long want of issue, else where barrennesse is the best blossome of beautie; or finallie, where pore men's children for want of their owne issue are not readie to be had'.[20] Even as recently as 1947, the novelist George Orwell attributed the English obsession with pets to 'the dwindled birthrate'.[21]

Once again, it is tempting to dismiss the use of pets as 'child substitutes' as a purely western phenomenon – perhaps something to do with the gradual disintegration of the traditional extended family[22] – but, as we have seen in the previous chapter, people in other cultures often take the childlike role of the pet to similar extremes. Juan and Ulloa refer to South American Indian women treating their pet poultry as if they were only children,[23] and several authors have described how the Australian Aborigines used to regard their dogs and dingoes with greater affection than they bestowed on their own offspring.[24] The Brazilian Kalapalo Indians not only treat their pets in exactly the same way that they treat their children, they even bury dead ones in precisely the same manner they bury infants that die during childbirth or before being named.[25] Several societies, including the land Dyaks of Borneo and a number of subarctic Indian groups also include pet animals within their system of 'teknonymy' – a system of naming in which parents are named after one of their children, rather than the western practice of naming children after their parents. Among these cultures, adults whose children have grown up and left home, or who have no children, are sometimes addressed by names which identify them as the parent of their favourite pet. The animal's name thus becomes part of the person's social identity, substituting for the name of the absent child.[26] Something similar to this exists in the West, when people occasionally append their own surname to that of their pet. Frequently, the parent–child aspect of pet-keeping is even more exaggerated in so-called 'primitive' cultures than it is in our own. If a woman in Britain or the United States were to breast-feed a puppy or a piglet in public (or even in private) she would probably be locked away for indecency. Yet in countless hunting and gathering or simple agricultural societies, the suckling of young animals is considered perfectly normal and natural (see chapter 4).

Even in cultures that do not ordinarily make pets of animals, their child-like qualities are often acknowledged. The Kenyah people of Borneo keep immense numbers of dogs. So many, in fact, that the dogs are a considerable public nuisance. The Kenyahs show little if any affection for these animals and discourage their children from playing with them. But, when asked why they do not kill them or get rid of them, one informant explained that they could not because the dogs 'are like children, and eat and sleep together with men in the same house'.[27] The Indians and Eskimos of Canada and Alaska have a distinctly utilitarian attitude to their sledge-dogs and rarely show them any affection. Nevertheless, childless couples will adopt a puppy and rear it in place of a child. These dogs are never harnessed or required to do any work. Instead they are given free run of the home, fed by hand and generally pampered.[28]

It is possible to infer the child-like status of many pets, not only from the ways in which people act towards them, but also from the animal's own physical appearance and behaviour. In the 12,000 or so years since its domestication, the dog has altered dramatically in size and shape, and some of these changes seem to reflect the human weakness for small, cuddly, infantile animals.[29] In many dog breeds there has been obvious selection for miniature size and this development has often been accompanied by the evolution of grossly foreshortened jaws and noses, relatively large and protuberant eyes and high foreheads. Pugs, bulldogs, Pekingese, King Charles spaniels, griffons, Lhasa apsos, Japanese chins, Chihuahuas, Boston terriers, Shih Tzus, Tibetan spaniels, affenpinschers, Maltese and Pomeranians are just some of the small, pug-faced or short-nosed breeds that fit into this category. All have been bred as house pets or lap-dogs, and all of them have comically appealing, anthropomorphic (or, more literally, paedo-morphic) facial expressions and a number of other physical characteristics which appear specially designed to pander to the human cute response. The flat-faced and fluffy Persian cat may well be a product of the same process.

And it is not only the shape of animals which has altered to suit our desire for things to mother. To varying degrees, the majority of domestic dog and cat breeds also exhibit behavioural 'neoteny' – that is, the retention of infantile or juvenile behaviour patterns into adulthood.[30] In one recent study at the University of Michigan, the development of wolf pups was compared with that of (superficially similar) Malamute puppies when both litters were reared by the same female wolf in two consecutive years. The Malamutes acquired adult body proportions more slowly, the development of their motor skills and physical coordination was retarded, and they showed a protracted period of dependence on (human) adults. In short, compared with wolf pups, the Malamutes were canine Peter Pans who never entirely grew up. The scientists who conducted this project also noticed that the Malamutes were more appealing to people. When some friends visited, the two daughters and the mother responded with compara-tively little enthusiasm when bottle-feeding the wolf pups, but the same family responded to the Malamutes with an 'orgy of maternalistic indulgence'. On its own, this was not surprising given the sort of bad publicity which wolves receive but, to everyone's astonishment, the wolf foster-mother behaved in exactly the same way. She washed the Malamutes earlier and more frequently than her own pups, she spent two to three times as many hours in the den with them, defended them more vigorously, played with them more, and showed more distress when one was removed.[31] This does not, of course, mean that the wolf was necessarily responding to the same cues as the people but the findings were, nevertheless, suggestive.

Taking the combined weight of all this evidence into account, it does seem hard to escape the conclusion that the child-like qualities of many animals can evoke in people a strong desire to nurture and provide parental care and protection. It also seems difficult to dispute the idea that certain people, such as nursing mothers or childless individuals and couples, succumb more easily and wholeheartedly to the infantile charms of such animals than others. If these assumptions are correct, then it is probable that over the centuries the 'cuter' more lovable animals – the ones that looked and behaved more like babies – were preferred and treated better than those which developed normally. This in turn will have produced strong selection in favour of characteristics, such as neoteny, which enhanced these animals' intrinsic appeal. The human cute response, or whatever we choose to call it, provides a plausible explanation for the origin of pet-keeping; this widespread human practice of adopting other species, forming affectional bonds with them, and treating them as members of the family. It does not, however, necessarily follow from this that pets are social parasites, or that pet-owners are being subtly manipulated, like the cuckoo's foster-parents, into providing attention and care which they would otherwise invest in their own offspring.

As with any analogy, it is possible to take the social parasite idea altogether too far. Evolution has forced the cuckoo to disguise its eggs accurately to resemble those of its host. If it did not the foster-parents would detect the interloper and either throw it out or abandon the nest. Even the cuckoo nestling is not entirely safe. Many songbirds are rarely if ever parasitized by cuckoos and this situation has almost certainly arisen because, in the past, they evolved the ability to discriminate between their own young and the parasite's. The

Oxford zoologists, Richard Dawkins and John Krebs, have referred to the relationship between the cuckoo and its hosts as an evolutionary 'arms race'; a perpetual cold war struggle in which the parasite busily refines its ability to deceive, while the host constantly improves its capacity to detect each new deception.[32] If birds are capable of evolving this level of discrimination, then so are humans, and pet-keeping would long since have ceased to exist. Puppies and kittens or Pekingese and pugdogs may be cute, but it is ludicrous to suggest that people are seriously incapable of distinguishing between them and real babies. It is equally absurd to argue that pet-owners are the victims of some elaborate subterfuge. Pets can be surprisingly clever at manipulating their owners' behaviour and emotions,[33] but more often than not, owners are perfectly well aware of what is happening, and are only too willing to oblige.

Journalists and social commentators frequently make sarcastic assertions about how the English or the Americans treat their pets as well as, if not better than, they treat their children. But this is simply untrue. We do not ordinarily abandon our children when we are tired of them or we want to go on holiday. We do not have them put down when they become unruly, we do not castrate them or sterilize them, or feed them on an uninterrupted diet of canned offal. Nor do we thrash them and rub their faces in it when they defecate in the house. Which ever way one looks at it, it is the pets who make the major sacrifices, not the owners. All those appealing deformities which we have bred into our pets cause many of them lives of continuous physical discomfort. The bulging eyes of the bulldog and the King Charles spaniel are subject to drying and are susceptible to injury. Their squashed up faces lead to respiratory problems and dental difficulties. The wrinkles and folds of skin around their faces which give them such comical and endearing expressions harbour bacteria which often give rise to serious infections.[34] None of this is consistent with the view of the pet as an accomplished social parasite. But, then, not all supposedly parasitic relationships are entirely what they seem.

At first sight, the life-cycle of the South American giant cowbird (*Scaphidura oryzivora*) seems very similar to that of the European cuckoo. It lays its eggs in the nests of other birds – especially a group known as oropendolas and caciques – and, if all goes according to plan, the host parents rear the cowbird nestlings as their own. The only obvious difference is that the baby cowbird does not kill or evict its foster-siblings. However, when examined in more detail, the story of the giant cowbird proves to be infinitely more complicated. In the first place, there are two quite distinct types of female cowbird: the 'mimics' and the 'dumpers'. Mimics are devious and stealthy; they lurk surreptitiously around the nest until the host birds are away and then steal in and lay single eggs which are disguised to look like those of the host. In fact, the mimics are just like cuckoos. The dumpers, on the other hand, are conspicuous and aggressive. They chase the future parents away and 'dump' a whole clutch of two to five undisguised eggs in the nest. The mysterious thing is that some oropendolas and caciques do not immediately evict the dumper's eggs, despite the fact that they are easily recognizable. As it turns out, they have a very good reason for not doing so.

Oropendolas and caciques are plagued by a particularly nasty kind of fly, the larvae of which literally eat their nestlings alive. To counteract this threat the birds adopt one of two strategies: either they build their nests near beehives, where the bees actively hunt down and kill the flies, or they allow their nests to be 'parasitized' by cowbirds. Nestling cowbirds spend a lot of time preening their nestmates to rid them of fly eggs and larvae. They also peck aggressively at any adult flies which enter the nest. In other words they protect the host's own offspring from fly infestation.

In areas where beehives are plentiful, oropendolas and caciques can and do discriminate between their own eggs and cowbird eggs. Only the disguised eggs of the stealthy mimics pass inspection. However, where beehives are in short supply, the hosts stop discriminating and allow dumpers ready access to their nests. What was once a relatively simple parasitic relationship, in which the cowbirds benefited at the expense of their hosts, has become a true symbiosis in which both host and parasite derive mutual benefit from coexistence and tolerance.[35]

The moral of this story is simply that there may be a great deal more to the owner–pet relationship than meets the eye. Pet-owners clearly do discriminate between pets and their own offspring. Yet they not only allow these animals to remain in the 'nest'; they actively seek out relationships with them and devote considerable emotional and financial efforts to their maintenance. Any biologist observing such a relationship between two species would be forced to conclude that the host, in this case *Homo sapiens*, was deriving some sort of hidden benefit from the presence of the parasite. The child-like qualities of pet animals probably do make them more lovable and endearing, but this fact on its own is inadequate to explain why we choose to share our lives, our homes, and our economic resources with companion animals. It seems far more plausible to suggest that this anomalous and unlikely partnership exists because people have something fairly substantial to gain from pet-ownership; something that is to some extent independent of economic considerations.

1, 2 'Before and after': modern agricultural intensification has drastically reduced the quality of life of many domestic animals.

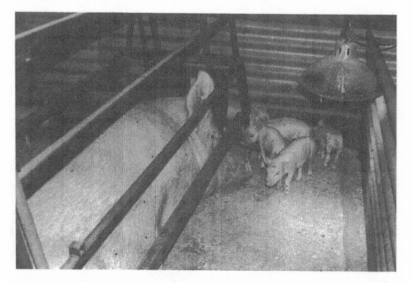

3 A few pet-owners are prepared to take their relationships with companion animals to bizarre extremes.

4 Hitler and Blondi. Affection for animals acquires a particular sinister quality when combined with misanthropic or xenophobic attitudes to other people.

5 M'tesa, despotic nineteenth-century King of Uganda, walking with his dog.

6 The family of Charles I, including Charles II (centre) and James II (to his right), a dynasty of dog-loving English monarchs.

7 Queen Elizabeth II and her corgis, maintaining the royal pet-keeping tradition.

8 Witches with familiars. During the sixteenth and seventeenth centuries, the possession of a pet could be taken as prima facie evidence of diabolical inclinations.

9, 10, 11 Like many other
hunting societies, the Barasana
Indians of Colombia keep pets
for pleasure rather than utility.

12 The 'cute response': the soft fur, large eyes, small noses and round faces of the puppy and kitten appear specially designed to appeal to human parental emotions.

13 The Punan Dyaks love their dogs, but not necessarily because they are economically useful.

14 The way in which people tend to respond to baby animals is superficially similar to the reaction of parent songbirds when confronted with a parasitic cuckoo nestling.

15 The first of William Hogarth's *Four Stages of Cruelty*. According to popular tradition, cruelty towards animals encouraged violence and sadism towards people.

16 Pets are now widely used in a variety of therapeutic contexts. Disabled and elderly people, in particular, derive considerable pleasure and comfort from the companionship of animals.

17, 18, 19, 20 Some of the ways in which dogs signal attachment and affection for their owners: relaxed physical intimacy; exuberant greeting; proximity and mutual eye-contact; undivided attention.

21 The rodeo, with its emphasis on gratuitous violence and domination, has been described as the modern equivalent of a public hanging.

22 The world's tropical forests are being destroyed at the rate of 150,000 square kilometres a year, victims of the ruthless expansionist philosophy that has dominated the recent history of our species.

PART III
An alternative view

CHAPTER 6

Pets as panacea

> He counteracts the powers of darkness by his electrical
> skin and glaring eyes. He counteracts the Devil, who is
> death, by brisking about the life.
>> Christopher Smart (1722–71), *My Cat, Jeoffrey*

In 1964 an American child psychiatrist, Boris Levinson, first coined the phrase 'pet therapy' to describe the use of pet animals in the treatment of psychiatric disorders. The idea that companion animals could be of practical value in health care came to him when he noticed that some of his severely withdrawn patients – children who had serious difficulty relating to or communicating with other people – seemed to have no problem striking up healthy friendships with his pet dog, Jingles. At first, the children incorporated Jingles into their fantasy play and completely ignored the attentions of the psychiatrist, but gradually, by carefully insinuating himself into these games, Levinson found that he was able to establish a rapport with the child and so begin therapy. In other words, the animal somehow acted as an 'ice-breaker', first helping to soften the child's initial hostility and reserve, and then providing a focus of communication between patient and therapist.[1]

Over the years which followed, Boris Levinson also made a variety of claims for the potential beneficial role of pets in human development. Among other things he argued that the experience of caring for a pet during childhood could make a person more sensitive to the feelings and attitudes of others; inculcate tolerance, self-acceptance and self-control, and provide an early introduction to the facts of life and death. He also believed that pets could enhance people's emotional development directly by acting as constant sources of companionship, comfort and security during periods of alienation, rejection or crisis. It was this unique capacity to offer unconditional and non-judgemental affection and support which Levinson saw as the key to the animal's importance as a therapeutic instrument. With careful application, he maintained, companion animals could be used as a deliberate form of therapy, particularly in institutions such as hospitals and nursing homes where people are necessarily separated from the support of relatives and friends.[2]

Levinson's early publications and reports were greeted by many of his colleagues with scepticism and a certain amount of derision, although he was by no means the first to attribute healing powers to animals, especially dogs. In ancient Egypt, dogs were sacred to the jackal or dog-headed god, Anubis, whose roles included physician and apothecary to the gods, and guardian of the mysteries of mummification and reincarnation. The dog was also the sacred emblem of the Sumerian goddess, Gula the 'Great Physician', and of the Babylonian and Chaldean deity, Marduk, another god of healing and reincarnation.[3] In ancient Greece, dogs played a central role in the cult of Asklepios (Aesculapius), the son of Apollo, who was known as the God of Medicine and the Divine Physician. Asklepios's shrine in the sacred grove at Epidaurus functioned as a kind of ancient health resort. It attracted crowds of suppliants seeking relief from a great variety of real and imagined maladies and, as part of the 'cure', provided one of the world's earliest recorded examples of institutionalized pet therapy. Treatment involved various rites of purification and sacrifice followed by periods of (drug-induced?) sleep within the main body of the shrine. During their slumbers the God visited each of his patients, sometimes in human form but more often in the guise of a snake or a dog which licked the patients on the relevant injured or ailing portions of their anatomy. In reality, the dogs that lived around the shrine were specially trained to lick people and make a fuss of them since it was believed that these animals were emissaries of the God and had the power to cure illness with their tongues. Inscribed tablets found within the precincts of the temple at Epidaurus testify to the miraculous powers of the local dogs:

Thuson of Hermione, a blind boy, had his eyes licked in the daytime by one of the dogs about the temple, and departed cured.

A dog cured a boy from Aigina. He had a growth on his neck. When he had come to the god, one of the sacred dogs healed him while he was awake with his tongue and made him well.[4]

The notion that dogs could heal injuries or sores by licking them persisted well into the Christian era. Saint Roch who, like Asklepios, was generally depicted in the company of a dog, seems to have been cured of plague sores by the licking of his canine companion. St Christopher, St Bernard and a number of other saints were also associated with dogs, and many of them had reputations as healers. France was the home of the extraordinary medieval cult of the greyhound saint, St Guinefort. According to legend, this faithful animal was killed by his noble master who, finding the dog drenched in blood beside his child's cradle, immediately assumed that it had devoured the infant. Only afterwards did he find the child sleeping peacefully, and the remains of a huge serpent torn to pieces by the dog's bites. Overcome with remorse, the knight threw the dog, Guinefort, into a well, covered it with a great pile of stones, and planted a grove of trees around it to commemorate the event. During the thirteenth century, this grove, about forty kilometres north of the city of Lyons, became the centre of a pagan healing cult. Peasants from miles around brought their sick and ailing children to the shrine where miraculous cures were performed. Needless to say, the official church took a dim view of such proceedings. A Dominican friar called Stephen of Bourbon had the dead dog disinterred, and the sacred grove cut down and burnt, along with the remains of St Guinefort. An edict was also passed making it a crime for anyone to visit the place in future. Despite these severe measures, vestiges of the cult were still in evidence as late as the nineteenth century.[5]

According to the Roman writer Pliny, it was not only the tongues of dogs that had healing properties. Devotees of the Gaulish goddess, Sequana, evidently pressed live puppies against their bodies in the belief that their ailments would be transmitted to the animal and so relieved.[6] Centuries later the close companionship of a 'Spaniel Gentle or Comforter' – a sort of nondescript, hairy lap-dog – was still being recommended to the ladies of Elizabethan England as a certain remedy for a variety of ills. William Harrison, in his *Description of England*, admitted to some scepticism on the subject:

> It is thought of some that it is verie wholesome for a weake stomach to beare such a dog in the bosome, as it is for him that hath the palsie to feele the dailie smell and savour of a fox. But how truelie this is affirmed let the learned judge.

But the learned Dr Caius was not beset with such doubts: 'though some suppose that such dogges are fyt for no service, I dare say, by their leaves, they be in a wrong boxe'.[7] He was of the opinion that a dog carried on the bosom of a diseased person absorbed the disease.

One of the earliest applications of animal therapy within an institution occurred in England in the late eighteenth century at a famous asylum known as the York Retreat. The scheme was the brainchild of William Tuke, a progressive Quaker. The York Retreat employed treatment methods which were exceptionally enlightened when compared with those which existed in other mental institutions of the day. Inmates were permitted to wear their own clothing, and they were encouraged to engage in handicrafts, to write, and to read books. They were also allowed to wander freely around the retreat's courtyards and gardens which contained various small domestic animals, such as rabbits and poultry, which patients were encouraged to care for and associate with. Tuke believed that the animals created a humanizing influence and argued that patients would 'learn self control by having dependent upon them creatures weaker than themselves'. A similar institution called Bethel, originally designed to house epileptics, was also founded in southern Germany towards the end of the nineteenth century. It is still in existence and currently boasts more than 5,000 inmates who suffer from a variety of mental and physical disorders. Therapy at Bethel includes two working farms, horse-riding, and large numbers of small companion animals.[8]

The early years of the twentieth century saw the appearance of the first scholarly papers on the potential psychological value of pet animals. In 1903, W. Fowler Bucke published a long and rather extraordinary analysis of 1,200 essays written by children about their pet dogs. In it, he noted the value of dogs as sources of affection when children feel lonely or unwell. He even speculated about the possible healing power of pets, given their capacity for relieving 'solitude'.[9] Sigmund Freud's classic case studies tended to emphasize the negative symbolic role of animals in the development of children's phobias and psychoneuroses,[10] but others saw a more positive side to the child–pet relationship. In a brief but insightful article published in 1944, the psychiatrist Bossard proposed that pets could enhance children's self-esteem, empathy and powers of communication, as well as encouraging them to come to terms with normal physical and bodily functions.[11]

Clearly, 'pet therapy' or at least the notion that animals can have a socializing or therapeutic influence was not new, but until Levinson came along it had been hardly more than an appealing idea with little sensible theoretical underpinning, and no scientific credibility whatsoever. Levinson's contribution was that he wrote seriously and extensively about the value of pets and suggested plausible and testable theories to account for the benefits of animal companionship. By doing so, he brought the whole phenomenon to the attention of scientists and health care professionals for the first time.

Among the first to put Levinson's novel theories to the test were a husband and wife team of psychiatrists at Ohio State University. Stimulated by the idea of using animals in medical institutions, Sam and Elizabeth Corson began looking into the feasibility of setting up what they called a 'pet-facilitated

psychotherapy' (PFP) programme within the psychiatric unit
where they worked. They selected fifty withdrawn and uncom-
municative patients most of whom had failed to respond
favourably to more conventional treatment methods, and
allowed each of them to choose a particular dog from kennels
adjacent to the hospital. The patients were then permitted to
interact with the chosen animal at appointed times each day.
Only three inmates failed to accept their chosen dogs and
withdrew from the study.

At the end of the study, the Corsons reported 'some
improvement' in all the remaining forty-seven patients
although, unfortunately, they gave details of only five subjects,
all of whom had improved markedly. Their overall assessment
of the value of PFP was, however, unequivocal. According to
their account, pet-facilitated therapy helped their patients to
develop self-respect, independence and self-confidence and
transformed them from 'irresponsible, dependent psychological invalids into self-respecting, responsible individuals'.
Contrary to expectation, patients did not become so attached
to their pets that they lost the will to interact with other
people. Rather, the dog acted as a social catalyst, forging
positive links between the subject and other patients and staff
on the ward, and thus creating a 'widening circle of warmth
and approval'. The Corsons believed that dogs were able to
induce such changes by providing patients with a special kind
of non-threatening, non-judgemental affection which, in their
own words, 'helped to break the vicious cycle of loneliness,
helplessness and social withdrawal'.[12]

This study was not without flaws. It depended entirely on
individual case histories and, for ethical and financial reasons,
no attempt was made to eliminate other forms of therapy
while patients were exposed to PFP. Also, there was no
matching control group – a similar group of patients which did
not receive PFP – to serve as a comparison. The Corsons,
however, did not regard these as serious defects because,
they said, they were not advocating PFP as 'a substitute for
other forms of therapy, but as an adjunct to facilitate the
resocialization process'. Later they went on to experiment

with animal therapy in a geriatric institution and again reported outstanding successes with some individual inmates; notably one old man who spoke a coherent sentence for the first time in twenty-six years. They also proposed the use of pet therapy in prisons, and this challenge has since been taken up by at least three major penal institutions in the USA.[13] The efficacy of such schemes remains to be established through careful research, but the organizer of one pet therapy programme at a maximum security mental hospital in Ohio has claimed a 50 per cent reduction in incidences of violence, attempted suicide, and in the need for tranquillizing medication.[14] Unfortunately, the data to substantiate these claims have never been published.

One of the better designed studies to date was carried out in 1981 by a team of consultant psychologists in Melbourne, Australia. This was an attempt to evaluate the beneficial influence of pet therapy on the overall morale and happiness of a group of elderly nursing-home residents. Fifty-eight patients occupying two wards of the Caulfield Hospital in Melbourne formed the main experimental population for this study. The majority were elderly (average age eighty years) and many were frail and suffered from various conditions typical of old people, such as arthritis, Parkinson's disease, senile dementia and cardiovascular ailments. At the outset, each patient was carefully assessed by the hospital staff on a variety of measures: e.g. happiness, mobility, alertness, relationships with other patients and staff, etc. A matching control group – that is, a similar group of elderly people occupying a separate ward – was also assessed in the same way.

At this point, a former guide-dog called Honey was introduced into the two experimental wards only. She remained there as a sort of resident hospital mascot for the next six months, during which time staff members repeated their psychological assessments of both experimental and control subjects at monthly intervals.

By the end of the study, a number of changes were observed in the experimental wards. Inmates were rated as happier, having more sense of humour, smiling and laughing more, being more alert and responsive, more easy going, more interested in others, enjoying life more, having a greater will to live, and having improved relationships with other patients and with staff. All of these changes were statistically significant and some, including alertness, will to live, and relationships with patients and staff, were highly significant. In contrast, the control group patients who had had no contact with the dog remained essentially unchanged over the same period. Moreover, at the end of the study they were found to be significantly less easy going, more withdrawn and less interested in others than members of the experimental group. They also spent more time alone or in bed.[15]

These findings need to be interpreted with caution. The apparent 'improvements' observed in the experimental groups were based on subjective assessments by members of hospital staff rather than on objective observations by independent investigators. The staff members involved were all in favour of the project and expected benefits, and it is therefore possible that they unconsciously biased their assessments in the expected direction. On top of this, we have the fact that the arrival of a dog in a geriatric ward is a distinctly novel event, and this novelty aspect may be sufficient on its own to generate transient improvements in patient condition.

These problems of interpretation are highlighted by a more recent effort to replicate and extend the Melbourne findings. The design of this later study was essentially similar – although it lacked a control group – but it included behavioural observations of patients and staff before, and at intervals after, the introduction of a therapeutic dog. As in Melbourne, both patients and staff felt that the dog had had a beneficial effect overall, although the staff were more favourable than the patients. The behavioural evidence, however, suggested that these effects were short lived. After an initial increase in activity and social interaction a short time after the dog's arrival, most patients gradually reverted to the more solitary behaviour they had shown before the study. Staff morale underwent a more dramatic and long-lasting improvement, and certain individual patients –

especially those who had been particularly withdrawn before-hand – seemed to benefit markedly.[16]

Various other studies have also explored the supposed therapeutic influence of companion animals. The majority have examined the impact of pets on disturbed, handicapped, depressed or elderly persons living within institutions, although a few have looked for health benefits of pet-ownership in the general population. Several of the clinical studies have reported striking improvements in the condition of individual patients, but the findings, overall, have been somewhat inconclusive and difficult to interpret.[17] Despite these reservations, therapeutic or remedial programmes involving animals have acquired a considerable following in recent years, particularly in the United States. A 1983 review of the literature on human–pet relationships, for example, listed thirty-four separate articles and publications dealing with the design or results of animal therapy programmes, and the numbers continue to grow each year.[18] Those who engage in or instigate such programmes are fired by the unshakeable conviction that certain people, particularly those in insti-tutions, can derive important psychological and medical benefits from the companionship of animals. Judging from the anecdotes and case histories which have emerged to date, the individuals who seem to benefit most are those who, for whatever reason, feel alienated or rejected.[19] Relationships with pets appear to be able to break down the barriers of despair and disillusionment which not only isolate such people from others, but also render them less accessible to more conventional forms of treatment.

Perhaps the most interesting and surprising evidence linking pets to human health came from a series of studies carried out at Maryland and Pennsylvania Universities. In 1977, Erika Friedmann, a graduate student at the University of Maryland, began a study of the effects of social conditions and social isolation on the survival of a group of heart-attack sufferers. Friedmann interviewed each of her subjects while they were still recovering in hospital; first, with a standard psychological mood test (since depression is known to be associated with increased vulnerability to heart disease), and, second, with an inventory of questions designed to explore every possible aspect of the patients' social life at home. Buried in the latter were some items on pet-ownership. Friedmann carefully followed up each of her subjects after they were discharged from hospital, and she found that after one year fourteen of the original group of ninety-two had died. She had all the data from her previous interviews and questionnaires on computer, so it was a relatively simple matter for her to ask the machine what significant differences there were between those who had survived the year and those who had not. As anticipated, certain types of social contact (or a lack of them) emerged as important predictors of survival but, contrary to expectation, the computer also told her that pet-owners had significantly better chances of surviving than non-owners.[20]

Not surprisingly, Friedmann and her advisors were sceptical about these initial results. They double-checked the original data for errors and, when they failed to find any, began looking for other plausible explanations. Some pets, for example, need regular exercise; could the results be merely a reflection of the additional physical exercise of walking dogs? They split the sample into dog-owners and other types of pet-owner and repeated the analysis, only to find that improved survival persisted in both groups. Then they tackled the possibility that pet-ownership was an effect of better health rather than a cause. For instance, it could be argued that healthier people – the sort of people who would tend to have less serious heart attacks – are also more likely to be pet-owners in the first place. Friedmann calculated an index of heart disease severity for each of her subjects but could find no correlation between lower initial severity and pet-ownership. Rather, it appeared that pets actually enhanced the recovery of their owners irrespective of the seriousness of the original attack.[21] They even countered the suggestion that pet-owners were psychologically different from non-owners and were therefore predisposed to survive better. In a separate survey of a random sample of pet-owners and non-owners they found no significant personality differences between the two groups.[22]

Having got this far, it seemed fair to assume that pets were, after all, having some positive influence on the survival of their owners. Admittedly, this influence was quite small – of the order of a 3 per cent decrease in the probability of dying – but in a country where more than a million people die each year from heart disease, this would still entail a possible 30,000 survivors in any given year. All that was needed now was some reasonable explanation to account for this extraordinary discovery.

Friedmann and her supervisors, Aaron Katcher and James Lynch, began looking for this explanation in people's physiological responses to animals. If it could be shown, for instance, that pets somehow helped their owners to relax physically then it would be possible to argue in favour of a direct therapeutic influence of pet-ownership. In the first of a series of experiments, they measured changes in people's blood pressure in four different situations: when they were sitting and resting quietly, when they were reading out loud, when they were talking to an experimenter, and when they were greeting their own pet dogs. At the time it was already well known that people find talking or reading out loud to another person stressful particularly if that person is someone in a position of authority. This effect was soon confirmed by subsequent blood pressure measurements. When Friedmann, Katcher and Lynch's subjects were reading out loud or talking to the experimenter their blood pressure was significantly elevated. But when the same subjects greeted and talked to their dogs, their blood pressure fell until it was equal to or slightly below resting levels.[23] At the time, this result led to some excited speculation about the possible calming effects of petting or stroking pet animals, but later work revealed that something far more subtle was involved.

It was already apparent from previous work that companion animals could make people feel calmer in strange situations. A psychology student based at the University of Lancaster in England had found that human subjects left to complete a test known as a 'manifest anxiety scale' with a dog for company displayed lower anxiety scores than those who completed the test alone.[24] Friedmann and her coworkers explored this phenomenon in more detail by examining the cardiovascular responses of children in two different experimental situations. The children in one group were brought into a room individually and introduced to an experimenter and his dog. They were then asked to rest quietly and then to read aloud while their blood pressure and heart rate were recorded at regular intervals by a machine. After these initial tests, the dog was taken out of the room and the same tests were repeated. Children from the second group went through the same procedure only in reverse. Initially they were tested in a room alone with the experimenter but then, half-way through, the dog was brought in and the tests repeated. In neither case were the subjects allowed to touch the dog or interact with it in any way. The results of this experiment clearly demonstrated that the children had no need to stroke the animal in order to be calmed by it. It was sufficient merely that it was present in the room. Whenever the dog was there, the children's resting and reading blood pressure was significantly lower. And, moreover, the children in the first group remained calm even when the dog was removed halfway through the test.[25] In other words, the dog's mysterious calming ability seemed to linger in the room even after the animal itself had departed.

To make matters even more complicated, it was later discovered that pet fish could also make people relax. In yet another experiment Aaron Katcher, a psychiatrist at the University of Pennsylvania, sat people down in front of an aquarium tank in his office and measured their blood pressure while they simply watched tropical fish. Once again, subjects' blood pressures dropped to levels lower than when they merely rested and contemplated a blank wall. The effect was also markedly stronger in hypertensive subjects than in those with normal blood pressure.[26]

The results of this early work on the immediate physical effects of animal contact also prompted a whole series of cross-sectional and prospective studies comparing the health status of large numbers of pet-owners and non-owners. Some of these surveys have failed to discover any significant difference in health between owners and non-owners,[27] and a few have detected apparent negative effects of pet-ownership. In one case, for instance, it was found that elderly women who reported low levels of attachment for their pets were actually worse off in terms of general happiness than those without pets,[28] while the results of a separate survey of young adults showed that the individuals who were most strongly attached to their pets were also more 'emotionally distressed' as well as being more vulnerable to life stresses.[29] The bulk of these surveys, however, have found some sort of association between pet-ownership and improved health status, particularly among people exposed to a recent traumatic event, such as the death of a spouse. In two separate studies of recently widowed women, pet-owners were found to be emotionally and physically more resilient to the stressful effects of bereavement.[30] Similarly, a prospective study of the effects of negative life events on the use of physician services by elderly people, found that pet ownership appeared to exert an ameliorating effect on the normally strong, positive relationship between life stress and poor health.[31]

Other studies have shed some light on the specific health problems which may be affected by pet ownership. According to the findings of a survey of some 5,741 people attending a free, cardiovascular 'risk assessment' clinic in Melbourne, Australia, pet-owners are significantly less at risk from heart disease. On average, the pet-owners in this study had significantly lower blood-pressure, and reduced concentrations of cholesterol and triglycerides in their blood compared with non-owners. Furthermore, these effects seemed to be independent of the type of pet owned, and stronger than the known effects of other positive lifestyle factors, such as regular exercise or low-fat diets.[32] Pets may also have a measurable effect on the prevalence of everyday, minor health problems. A prospective study in England that measured people's health and overall well-being before, and at intervals after, the acquisition of a new pet, found reductions in minor health problems, such as colds, coughs, headaches, backaches and insomnia, which lasted for up to ten months after pet acquisition, especially among dog-owners. Pet-owners also reported improved psychological well-being and self-esteem scores compared with a group of non-owners who were assessed over an equivalent period of time.[33] As yet, it is too early to say whether all of these apparent health effects are the direct result of pet-ownership. It is possible, for example, that pet-keeping and improved health are each independently related to some other, as yet unidentified, factor or factors. Nevertheless, the evidence which has now accumulated is certainly sufficiently compelling to merit further, more detailed scientific investigation.

Judging from this evidence, animals seem to be able to influence us beneficially in a variety of distinct (though not mutually exclusive) ways. At the most basic level they can help us relax merely by being diverting. There is nothing new or profound in this. It has been known since the 1950s that any stimulus which is attractive or which concentrates the attention has a calming effect on the body. An aquarium displaying the scintillating movements of jewel-like, tropical fish is an entrancing visual spectacle which can induce an almost hypnotic state of relaxation. Log-fires and kinetic sculpture are comparable, though, perhaps less engrossing, examples of exactly the same sort of thing.[34] Distracting and absorbing images may be fairly commonplace in our lives, but it would be wrong to belittle the potential influence of such phenomena on our physical health. A recent detailed review on the subject suggests that distracting stimuli may play an important role in helping people to cope with pain and many other forms of distress.[35]

Animals also appear to be able to make strange situations and people less alarming. Both Levinson and the Corsons observed the ability of animals to, so to speak, 'break the ice' in encounters between patients and their therapists, and this response appears to be associated with physiological changes. Children introduced into a strange experimental situation are understandably alarmed and reveal their anxiety through their elevated blood pressure. Add an ordinary domestic dog to the same situation, however, and the effect on the children is immediate. Suddenly, the unknown experimenter and the strange circumstances become less menacing and they remain so even after the dog itself has been removed. The reassuring properties of pet animals have been confirmed separately by the findings of a New York psychologist, Randall Lockwood. Lockwood was interested in how the presence or absence of pet animals affected people's perceptions of social encounters and the individuals involved in those encounters. To measure this, he subjected a sample of college students to a procedure known as the 'Thematic Apperception Test' or TAT for short. The TAT consists of a series of simple drawings depicting individuals or groups of individuals engaged in various ambiguous interactions. Subjects are asked to describe the mood of the scene and the individuals taking part in it using a selection of so-called 'bipolar' adjectives (e.g. happy–sad, intelligent–unintelligent, dangerous–safe, etc.). Lockwood tested his subjects with a range of either standard TAT pictures or the same pictures after the inclusion of an animal, such as a dog or a cat. Almost without exception he found that situations involving animals were perceived in a more positive light, and that people with animals were seen as friendlier, happier, less tense and less of a threat to others.[36]

Politicians, of course, have been aware of this for some time, and frequently enhance their own public image by appearing with their pets. Franklin Delano Roosevelt used his dog Fala to ingratiate himself with America's dog-lovers. Richard Nixon's dog, Checkers, saved his campaign for the vice-presidency, and considerably prolonged his political career, and Lyndon Johnson was deluged with protest mail when he was filmed lifting beagles up by the ears.[37] In the 1980 congressional elections in the United States, Republican candidates were advised to have themselves photographed with pets in order to make them more appealing to the electorate.[38] Millie, the springer spaniel belonging to George and Barbara Bush, had her own fan club, and a book published in her name called *Millie's Book* which earned hundreds of thousands of dollars in royalties. Millie was also used to score political points against Bush's rivals. Referring to Bill Clinton and Al Gore, Bush stated publicly during a campaign address that Millie 'knows more about foreign policy than these two bozos'. Not to be outdone, the Clintons' cat, Socks, now has his own fan club which receives more than 200 letters a day, and has its own permanent director and staff who answer the mail and publish Socks's monthly fan club newsletter.[39]

This ability of animals to make their owners seem friendlier and less threatening may help pet-owners socially. Many pet-owners claim that their animals have increased their circle of acquaintances and helped to make them new friends.[40] An English zoologist called Peter Messent was the first to test this claim by measuring the amount of positive social interaction engaged in or received by people walking their dogs in public parks and gardens. Comparing the figures he obtained with those based on observations of people strolling through the same parks either on their own or with small children or babies, Messent found that, on average, dog-walkers had a greater number of positive interactions, and more of these interactions involved extended conversations.[41] This phenomenon may be of particular importance to people suffering from physical disabilities. Able-bodied persons frequently experience awkwardness, aversion or guilt when confronted by disabled individuals, and show a marked tendency to shy away from them. Such habits reinforce the disabled person's feelings of social alienation and rejection, and are difficult to overcome. Studies in California, however, have shown that wheelchair-bound adults and children experience a dramatic increase in positive social interaction and acknowledgement from passers-by when accompanied by their service dogs.[42]

The tendency for people to perceive pets and their owners in a positive light is unlikely to be universal. An orthodox Moslem, for example, might view a man accompanied by a dog less positively than, say, a European or North American. In other words, the reassuring image of the pet is a cultural symbol of safety and security, rather like a British bobby on a bicycle or an elderly gentleman smoking a pipe. But this still raises the question why, within our culture, companion animals have acquired this symbolic connotation. In many societies a man's personality or temperament may be assessed from the condition of his livestock. Among the Nunamiut Eskimos, for example, a hunter's character will often be judged by the mood of his dog-team. If the dogs are surly and bad-tempered, the hunter will be judged as aggressive and unpredictable, and vice versa if they are relaxed and good-natured.[43] So it is not the animal itself which is disturbing or calming so much as the way in which it reflects the personality of its owner. Thus the happy, friendly dog is seen as an extension of a basically happy and friendly owner. Had Randall Lockwood included drawings of a cowering mongrel or a snarling Rottweiler in his TAT study, doubtless the scenes would have been construed differently.

According to popular tradition, people who are cruel to animals are more likely to behave sadistically toward other people. Thomas Aquinas (1225–74), for example, ruled that cruelty to animals could have a brutalizing influence on men, and dispose them to inflict cruelty on fellow humans. In his *Four Stages of Cruelty*, the painter William Hogarth (1697–1764) depicted the exploits of Tom Nero, torturing animals as a child, flogging an injured horse as a young man, before murdering a prostitute and, eventually, finishing up as a cadaver on the anatomist's slab. Like Aquinas, the eighteenth-century German philosopher Immanuel Kant also argued that 'he who is cruel to animals becomes hard also in his dealings with men', and more recently, Mahatma Gandhi suggested that the moral fibre of a nation could be judged by its treatment of animals.[44] Modern press reports lend further support to this apparent association between cruelty to animals and violence toward people. The notorious 'Son of Sam' murderer, for example, was reputed to have hated dogs, and to have killed several in the neighbourhood where he lived. Albert DeSalvo 'the Boston Strangler' trapped dogs and cats, placed them in orange crates, and then shot arrows through them, and a young murderer in New York who recently admitted to killing for fun, amused himself by pouring ammonia in fish

tanks and torturing animals as a child. A recent study of
criminal psychopaths in the USA has even demonstrated a
statistical correlation between the two tendencies. During
interviews, a significantly higher proportion of psychopaths
reported inflicting cruelty on animals during childhood than a
matching control group of convicted felons.[45] Taking all this
evidence into consideration, it appears that people may after
all have some justification for regarding the owner of a healthy
and contented pet as a safer and more appealing person.

Finally, the ideas and observations of Levinson, the Corsons,
and many others suggest that the most important therapeutic
benefits to be gained from companion animals derive from the
particular kinds of relationships we have with them. Attempts
to demonstrate these benefits using pet-therapy have
produced somewhat equivocal results,[46] but this is not
altogether surprising. Not long ago, a scientific review of
over 500 studies involving different kinds of conventional
psychotherapy concluded that none of them provided good
evidence that psychotherapy was any more effective than
treatment with placebos.[47] The problem is not so much that
the therapeutic techniques are necessarily ineffective, but
rather that it is virtually impossible to separate the specific
effects of treatment from the more general emotional benefits
people derive from simply having someone listen, in an
uncritical and sympathetic way, to their problems. And here,
perhaps, lies the key to understanding the potential psychological rewards of pet-ownership.

Sitting in front of an aquarium and meditating on its
contents is certainly relaxing, but few pet-owners regard their
pets as passive objects designed purely for distraction and
entertainment. Similarly, dogs and cats may be comforting
cultural symbols of harmless respectability, but few people
would admit to keeping them for this reason alone. In every
single survey of pet-ownership which has ever been conducted,
the vast majority of owners have given 'companionship' or
'friendship' as their principal reason for keeping a pet.[48] Pet-
owners do not value their animals primarily as objects, but
rather as subjects; as distinctive personalities with whom they

have affectionate relationships not dissimilar to the kinds of affectionate relationships they have with close friends and relatives. In other words, if we wish to understand the possible contribution of pets to human health and welfare, then surely we should begin by exploring the influence of close relationships, in general, on our mental and physical condition.

CHAPTER 7

Health and friendship

A faithful friend is the medicine of life.

Ecclesiasticus 6: 16

Humans are intensely sociable and gregarious creatures, probably more so than any other living mammal. Of course, people choose to associate with each other for a variety of reasons, not all of which are particularly sociable. If I wish, for example, to gain access to certain books, I will need to join a library and interact and cooperate with librarians. I may find these interactions rewarding in themselves but, more likely, I will view them simply as a means to an end; a purely practical method of attaining my own selfish, literary objectives. No offence to librarians intended, but I might equally interact with a dog in order to, say, train it to fetch my slippers every morning. On the other hand, I have also, like the majority of humans, devoted a considerable portion of my life to developing and maintaining close social relationships, and I have done this purely for its own sake without any obvious ulterior motives in mind. I do not value my friends for the services they may perform for me in the future; I value them for their friendship *per se*. Likewise, I do not value my pet cat because he is in any sense useful; I merely appreciate his company. This tendency of humans to value companionship and friendship for its own sake has been given a number of different labels – social motivation, affiliation, bonding, etc. – depending on the context in which it occurs, but all of these terms share in common the idea that people need and derive something special from social interaction and support.

Something above and beyond the mere gratification of short-term, material goals.[1]

From a biological perspective, this social need – our apparent dependence on others – can be viewed as the outcome of millions of years of evolutionary adaptation to living in complex, highly organized societies. It is fairly safe to assume that humans evolved in this direction because individuals who cooperated with each other in organized social groups did better, overall, in the lottery of natural selection than those who struck out on their own. There are, after all, a great many advantages to living in groups: there is both safety and strength in numbers, and groups can engage in cooperative ventures such as big game hunting or agriculture which are beyond the scope of most individuals.[2] One could even argue that group life is essential to human survival since most of the knowledge and skills people need in order to exist must be acquired, through teaching and imitation, from other individuals. In short, the cumulative benefits of sociability are so overwhelming that group affiliation, social bonding – call it what you like – has become an end in itself and not merely a means to an end.[3] It has acquired many of the properties of a physical need such as hunger; a need which must be satisfied before we can feel entirely happy and fulfilled as individuals.

People generally know soon enough when they are dehydrated, in need of food, or exposed to some kind of noxious stimulus. They start experiencing unpleasant subjective feelings such as thirst, hunger and pain which goad them into taking appropriate corrective action. Messages of this kind have also evolved to notify us when we are separated from and in need of close, affectionate contact with others. These messages are often collectively referred to as 'loneliness'. Loneliness is something that nearly everyone experiences at some stage during their lives, and it can be precipitated by almost any event involving a change in the quality or number of one's social relationships. The death of a close friend or relative, or a divorce or separation, are extreme examples, although even relatively minor upheavals can also create profound distress. Moving to a new neighbourhood, a new school or a new job; even a change in status such as a promotion at work can result in moderate to severe bouts of loneliness.[4] People vary markedly in how susceptible they are to the feelings of isolation associated with loneliness. When human subjects have been isolated in featureless rooms as part of psychological experiments, some have managed to remain for periods of up to eight days without feeling anything more than slight nervousness or unease. Others have been ready to batter the door down within a few hours.[5] Loneliness must also be clearly distinguished from the state of being alone. The latter may be actively sought out and enjoyed. The former is invariably a negative concept.[6] Hence, it is possible for a person to be alone yet contented hundreds of miles from the nearest human being, while the same individual may feel desolate with loneliness in the middle of a large crowd of people. In this respect, loneliness differs from the symptoms of physical deprivation. Human beings require absolute amounts of food and water in order to remain reasonably healthy. But the optimum amount of social contact a person needs will vary enormously according to his or her state of mind at the time.

Loneliness is regarded as such a painful and unpleasant sensation that, since time immemorial, societies have used solitary confinement, exile and social ostracism as methods of punishment. The autobiographical accounts of religious hermits, castaways and prisoners of war give us a clear picture of the psychological effects of enforced social isolation. Most describe feelings equivalent to physical torture which increase gradually to a peak before declining, often quite sharply. This decrease in pain is generally associated with the onset of a state of apathy and despair, sometimes so severe that it involves complete catatonic withdrawal.[7]

These symptoms bear a striking resemblance to those exhibited by young children when separated from their mothers. Initially, the abandoned child becomes literally frantic with distress; crying and screaming loudly, throwing himself about and shaking his cot with frustration. Eventually however, within hours or sometimes days, the outward signs of distress diminish and he becomes apathetic and withdrawn. In the words of the psychiatrist, John Bowlby, the child at this stage 'is in a state of unutterable misery'.[8] To the psycho-analyst this resemblance between adult loneliness and childhood separation distress is hardly surprising, since psychoanalytic theory assumes that adult social behaviour and needs are shaped, to a large extent, by early relationships with parents and other care-givers. To a degree this idea is borne out by the results of research. For example, it has been found that people whose parents divorced when they were children are more vulnerable to adult loneliness than those brought up within intact families.[9]

The unpleasant symptoms of loneliness and the pleasurable sensations we derive from close companionship and social support appear to be mediated by biochemical processes in the brain. Precisely how they are mediated is currently the subject of intense scientific debate, although the results of some studies strongly incriminate a group of neurochemicals known as endogenous opioids. In terms of their structure and effects, endogenous opioids resemble the opium derivatives, morphine and heroin. In the brain they serve a huge variety of functions, including those of natural sedatives and painkillers. In 1980, however, a group of psychologists in Ohio suggested that endogenous opioids could also play a part in 'modulating social emotions and behaviour'. In a series of experiments, they briefly isolated young animals such as chicks and puppies from their mothers and siblings. Predictably, the babies reacted to this sudden separation by uttering piercing distress calls and by trying to regain contact either with their families or a substitute object such as an experimenter. However, when very low doses of morphine were administered to these animals, the frequency and intensity of distress calling dropped dramatically and they showed much less interest in restoring social contact. This effect could be reversed by giving the animals another drug, naloxone, which blocks the action of opioids on the nervous system. Although the results obtained from these experiments are open to a variety of interpretations, the authors argue that their findings are consistent with the idea that endogenous opioid secretion contributes to the sense of security and well-being which young animals appear to derive from close social contact. They also suggest that when the animal is involuntarily separated from its parent or litter-mates, opioid secretion declines and it experiences unpleasant sensations analogous to drug withdrawal which, in turn, stimulate it to re-establish contact. The authors also imply that brains become more tolerant of these effects with age. Hence the fact that infantile attachments and separation distress are generally far more extreme than those of adults.[10]

As well as affecting young animals' social attachments, opioids may also mediate mothers' responses to their infants. Endogenous opioid secretion has been found to be closely involved with the induction of maternal behaviour in sheep in the period immediately after giving birth. Ewes treated with the opioid blocker, naloxone, at the time of birth fail to bond with their lambs, and appear to regard them with indifference or hostility. In rhesus macaques (*Macaca mulatta*), the mothers of four- to ten-week-old infants were also found to groom their infants less often, and show less protective behaviour towards them, when administered low doses of naloxone.[11] Positive social interactions among adult monkeys appear to involve similar neurochemical processes. Previous studies of social behaviour in non-human primates have demonstrated the importance of grooming as a means of establishing and maintaining affiliative relationships.[12] When researchers in Cambridge examined the relationship between grooming interactions and brain opioids in talapoin monkeys (*Cercopithecus talapoin*), they found a strong positive link between the two. Opioid levels in the monkeys' cerebrospinal fluid rose when they were being groomed by conspecifics but fell again when grooming stopped. Similarly, naloxone administration increased monkeys' motivation to be groomed while morphine administration decreased it. The authors of this study conclude that their findings 'support the view that brain opioids play an important role in mediating social attachment and may provide the neural basis on which primate sociality has evolved'.[13]

It is by no means certain that human social motives are mediated in similar ways, but there is at least some circumstantial evidence linking opioids to social behaviour. For some time it has been recognized that the typical symptoms of human loneliness and social isolation – the initial agony and the protracted feelings of hopelessness and despair – closely resemble those experienced by people suffering from narcotic withdrawal.[14] Human social needs, and the degree to which people become dependent on opiate drugs such as heroin are also strongly influenced by mood and circumstances, and in surprisingly similar ways. The severity of a drug-addict's withdrawal symptoms, for example, will generally depend on how much of the drug he has been taking and for how long.[15] Similarly, the degree of distress and unhappiness caused by social separation and loss tends to depend on the intensity and longevity of the relationship that has been lost or severed.[16] Also, people seem to find it easier to cope with solitude when they themselves are feeling happy and confident, and the same appears to be true of drug abuse. Thousands of American soldiers resorted to regular heroin use as a means of coping with the horrors of Vietnam but, contrary to expectations, relatively few relapsed into addiction on returning to the comparative peace and tranquillity of the United States. Likewise, many people are prescribed habit-forming doses of painkilling opiates in hospital, but few exhibit serious withdrawal symptoms once discharged.[17] Finally, it has been suggested that typical addicts may possess personality disorders that predispose them to drug-dependence. Studies have described them as sociopathic; as exhibiting an inability to establish or maintain close relationships, and possessing a self-centred, superficial attitude to others, combined with a low self-esteem and a belief that others dislike them.[18] These may of course be effects rather than causes of addiction but, nevertheless, they have led some authorities to conclude that the addict is someone who turns to heroin in order to compensate for a deficit of affection and companionship.*

Humans appear to have evolved these painful and distressing methods of alerting themselves to social deficits for a very good reason. A growing body of scientific evidence now suggests that loneliness and social isolation can seriously damage our health. Among children, for instance, loneliness has been linked with poor grades, expulsion from school, truancy, and engaging in delinquent acts such as theft, gambling and vandalism.[19] It has also been found that people with few friends or those who are poor at making friends are, statistically speaking, more vulnerable to a variety of both trivial and serious medical conditions.[20] A recent review of the medical literature on the relationship between coronary heart disease and social support provides overwhelming evidence, for example, that less socially integrated people are between two and four times more likely to die of heart disease than those who are better integrated socially.[21]

Being single, as opposed to married, also appears to constitute a significant health risk. Hospital-based studies in which background information was obtained on people suffering or dying from known causes have revealed that single people, irrespective of their race or age, are far more likely to suffer or die from strokes, heart disease, rheumatic fever, pneumonia, diabetes, nephritis, tuberculosis and most forms of cancer than their married compatriots. They are also more likely to be institutionalized (for any reason), more likely to suffer from severe psychiatric disorders, more likely to be involved in automobile and other accidents, and more likely to commit suicide. A lot of these associations are stronger in men than in women. For example, between the ages of twenty-five and fifty-five, the death rate from heart disease for unmarried, non-white American males is roughly double that of their married counterparts. Likewise, unmarried male smokers of more than twenty cigarettes per day are, on average, between

---

* Such ideas, of course, strongly echo what people have also said about the supposed emotional shortcomings of pet-owners (see chapter 2).

60 and 70 per cent more likely to die an early death than married, male smokers of equivalent age.[22]

In addition to apparently increasing our initial risks of succumbing to illness, loneliness also seems to be a barrier to recovery. Heart-attack survivors who are socially isolated – who have few friends, family members or positive social contacts – are roughly 50 per cent more likely to die within three years of their attack than those with high-quality social lives. Similarly, a study of over 27,000 cancer patients found a 23 per cent higher mortality rate among those who were unmarried. People with terminal illnesses also appear to die more quickly if they have few friends or visitors than if they are popular and have large supportive families.[23]

An interesting triangular relationship also seems to exist between social support, depression, and vulnerability to disease. A variety of studies have shown that depression can increase a person's risk for coronary heart disease, as well as having a direct suppressive effect on the immune system – the system responsible for protecting the body from invasion by bacteria, viruses and other disease-causing agents such as tumour cells. An even larger number of studies, however, have demonstrated an inverse association between depression and social support. In short, people with supportive social networks are much less likely to become depressed, and are therefore less susceptible to medical problems.[24]

Of course, some of these findings need to be interpreted with caution. One of the problems with so-called 'retro-spective' or correlational studies – studies which, after the event, look back at a patient's history or lifestyle – is that they are unable to separate out causes and effects. It could be argued, for example, that people with mental disorders or poor physical condition are less able to establish and maintain stable interpersonal relationships than those with more robust constitutions, in which case loneliness would be a consequence of ill-health rather than a cause. However, the results of several large-scale 'prospective' (i.e. before and after) studies conducted in the United States, Scandinavia and elsewhere point strongly and consistently in only one direction. People who report the lowest levels of social integration and support at the beginning of these studies are usually about twice as likely to die in subsequent years than those with the richest social lives. In other words, illness seems to be a consequence of social isolation rather than a cause. Furthermore these effects of social isolation occur independent of other socioeconomic and demographic factors, and hold true for a wide range of diseases, including coronary heart disease and many forms of cancer.[25]

Although the relationship between psychosocial factors and physical health is now well established, the precise chain of neural, hormonal and biochemical events involved in this dialogue between mind and body remains the subject of intense speculation. Nevertheless, one factor which seems to lie at the heart of the whole process is the phenomenon of stress. Since the publication of Hans Selye's book *The Stress of Life* in 1956, it has become increasingly obvious that unpleasant, aversive or *stressful* life events and experiences can exert a decidedly negative influence on both our emotional state and our health.[26] The processes underlying this so-called 'stress response' are now reasonably well understood: when people and other animals are subjected to unpleasant or painful stimuli, their bodies respond by secreting a group of hormones from the pituitary and adrenal glands. These hormones ordinarily serve to prepare the body for so-called 'fight or flight' reactions, and, once the emergency is over, hormone secretion generally declines to normal. However, under certain circumstances, particularly when the source of pain, anxiety or distress cannot be readily avoided or controlled, the stress response of the pituitary–adrenal system may become prolonged and exaggerated, thereby producing a number of deleterious consequences. These include interfering with the action of insulin; causing loss of calcium from bones; suppressing growth; and inducing excessive gastric secretion, menstrual irregularities and high blood pressure. Most important of all, prolonged stress causes suppression of the immune system.[27] The potential medical consequences of chronic exposure to large amounts of recent life stress are, thus, fairly predictable: increased danger of diabetes and stomach ulceration, increased susceptibility to infectious diseases and cancer, more allergies, greater risks of coronary heart disease and cardiac failure, and overall degeneration of health.[28] It is also acknowledged that chronic stress is a major causal factor in the onset of depression and other mental disorders.[29]

Exposure to stress also affects human behaviour. In particular, people tend to seek each other out during stressful or troubled times. Threatening situations such as war, natural disasters and other catastrophes are well known for bringing people together and promoting a sense of unity, solidarity and *esprit de corps*. This phenomenon has even been demonstrated in the laboratory. When psychologists have tried threatening human volunteers with either 'mild' or 'painful' electric shocks, and have then offered them the choice of waiting alone or in the company of other subjects, those faced with the painful shocks showed a much stronger preference for companionship than those threatened with the mild ones. Stanley Schachter, the author of this experiment, concluded that 'the state of anxiety leads to the arrival of affiliative tendencies' and that people serve 'a direct anxiety reducing function for one another, they comfort and support, they reassure one another and attempt to bolster courage'.[30] There is equally little doubt that this beneficial role of social companionship under stress can have far-reaching medical consequences. According to a report published by the US Surgeon General's Office in 1973:

the most significant contribution of World War II military psychiatry was recognition of the sustaining influence of the small combat group or particular members thereof, variously termed 'group identification', 'group cohesiveness', 'the buddy system' and 'leadership'. This was also operative in noncombat situations. Repeated observations indicated that the absence of such sustaining influences or their disruption during combat was mainly responsible for psychiatric breakdown in battle. These group or relationship phenomena explained marked differences in the psychiatric casualty rates of various units who were exposed to a similar intensity of battle stress.[31]

Similarly, during interviews, many survivors of the Nazi concentration camps admitted that the strength they were able to derive from the company of their fellow prisoners was by far the most important factor contributing to their will to live. Victims who were separated from their loved ones often gave way to profound despair, lost hope and perished.[32] The situation is more or less summed up by the words of the American psychologist, James Lynch:

Those individuals who lack the comfort of another human being may very well lack one of nature's most powerful antidotes to stress. Thus, individuals who live alone – widows and widowers, divorcees, and single people – may be particularly vulnerable to stress and anxiety. The presence of a friend or companion may not only help suppress fear and physical pain, but it may also reduce the 'wear and tear' on the heart that occurs under stress and chronic anxiety. The increases in cardiac death rates for those who live alone may be due, in part, to the fact that these individuals continuously lack the tranquillizing influence of human companionship during life's stresses. Tranquillizing drugs may not be able to fill the void, and in the final analysis they may not be anywhere near as effective as the calming capacity of human friendship.[33]

It appears then that, by protecting or buffering us from the everyday 'slings and arrows of outrageous fortune', our social relationships play a crucial role in maintaining our health. It should be emphasized, however, that not all social partners are equally good at providing this service. Social relationships can *cause* stress as well as relieving it. For instance, in a five-year, prospective study of the incidence of heart disease in 10,000 Israeli men, it was found that, among the men with high anxiety levels, those with 'loving and supportive wives' were about half as likely to develop heart conditions during the period of study compared with those with less affectionate spouses.[34] Conversely, research at Ohio State University has demonstrated a clear connection between marital conflict, depression and immunosuppression. People in poor-quality marriages are apparently in worse shape mentally and physically, and those couples who display more hostile behaviour towards each other also tend to attain lower scores on various different measures of immune function.[35] Deleterious effects of marital conflict may also help to explain the higher prevalence of illness and mortality among recently divorced persons. Never married or widowed, white American males, for example, are between 45 and 55 per cent more likely to die of coronary heart disease than married males of the equivalent age. But, among those who are divorced, the likelihood increases to over 100 per cent. Separated and divorced people are also greatly over-represented in psychiatric hospitals compared with other categories of single people.[36] Finally, our social partners can also have a profoundly damaging effect on our health by simply abandoning us for one reason or another. Most authorities on the subject of stress now agree that disruptions or disturbances in interpersonal relationships – divorce, separation, rejection by a loved one; the death or loss of a spouse, a parent, a child, a sibling or a close friend – are among the most stressful events a human being can experience during his or her lifetime.[37]

Humans, of course, do not have a monopoly on the protective and restorative effects of social support. Many other social animals show exactly the same kinds of responses to isolation and companionship as humans. They also show a tendency to decline physically and psychologically when socially isolated. Infant monkeys separated from their mothers, and adult monkeys isolated from their peers develop the characteristic symptoms of stress, including stomach ulceration and immunosuppression.[38] Non-human primates also appear to show the same social responses to stress, and with similar beneficial consequences. In a study published in 1992, for example, forty-three male cynomolgus monkeys (*Macaca fascicularis*) were divided into two groups and observed and monitored for immune function over a period of two years. The members of one group were allowed to remain with the same social partners for the entire period, while those in the second group were assigned new partners each month. This latter, 'unstable' group condition was already known to be stressful for the monkeys and, not surprisingly, the animals in this group showed reduced immune function compared with those in the 'stable' group. More interestingly, however, monkeys in the unstable group showed higher levels of affiliative behaviour – i.e. spent more time in contact with or grooming each other – and the most affiliative animals showed enhanced immune responses compared with their less affiliative fellows. It appears that chronic social stress, somewhat paradoxically, increased these monkeys' social motivations, and that positive social contact helped some of them to cope with the potentially pathogenic effects of socially induced stress.[39]

Affectionate social contact also seems to enhance some animals' ability to cope with harmful lifestyle factors. In an enlightened experiment conducted at Ohio State University in 1980, researchers examined the effect of positive social interactions on the development of coronary heart disease in laboratory rabbits. (The study was enlightened because they chose to pamper their experimental animals rather than stressing them as is usually the case.) In the study, two similar groups of rabbits were housed in identical laboratory conditions. One group was given no special treatment and the other was visited several times a day and stroked, talked to and played with on a regular basis. The rabbits belonging to the latter group soon learnt to recognize the experimenter and eagerly sought her attention. From the first day of the experiment both groups were fed on a diet rich in cholesterol, a diet which ordinarily would predispose them to hardening of the arteries. Six weeks later the rabbits were killed humanely and their aortas (the main blood vessel rising from the heart) were examined for damage. It soon emerged that those which had been handled and played with had 60 per cent less damage, on average, than those which received the standard laboratory treatment.[40]

In another experiment, domestic chickens were divided into three separate groups within a day or two of hatching. Over the next four weeks one group was ignored (i.e. given minimal human contact), the second group was 'hassled' (i.e. shouted at, subjected to loud noises, etc.), while the third group was 'socialized' or tamed by treating them gently, talking to them and hand-feeding them. All three groups were then inoculated, first with pathogenic bacteria and then with foreign blood cells, to test the effectiveness of their immune responses. The tamed birds showed significantly more pronounced immunity to both substances than the chickens from either of the other two groups. In both this experiment and the previous one involving rabbits, positive social interactions with people somehow rendered these animals more resistant to the deleterious effects of two quite different categories of pathogenic agent.[41]

During the 1970s another series of interesting, though ethically disturbing, experiments involving dogs demonstrated profound calming effects of companionship in stressful situations. To cut a long and distinctly unpleasant story short, researchers at Johns Hopkins Medical School measured increases in heart rate and blood pressure in dogs when they were subjected to painful electric shocks. Predictably all the dogs showed large and consistent increases in cardiac response when shocked, but only when left on their own. When a human experimenter sat with the dogs and petted them, they seemed to find the pain of the shocks less severe. In fact, their heart rate and blood-pressure increases during petting were roughly half as great as those obtained when the dogs were isolated. In later experiments the dogs were trained to anticipate the shocks by preceding them with an audible tone. Following training, the dogs revealed their anxiety by exhibiting an immediate 50–100 beat-per-minute acceleration in heart rate whenever the tone sounded. But, once again, the effect was strongly modified by the presence of a person. When an experimenter sat with the dogs and petted them while the tone and shock were delivered, the usual marked increase in heart rate was either eliminated altogether or changed to a decrease in heart rate.[42] Positive social contact with a person had such a profound influence on these dogs that a situation which, in one case, was extremely upsetting and stressful became hardly stressful at all.

It is evident from these kinds of results that animals can also be calmed and reassured by close relationships with other individuals. More to the point, perhaps, those individuals do not necessarily need to belong to the same species. The remarkable, stress-reducing or stress-buffering effects of social relationships and attachments appear, under certain circumstances, to transcend the barriers that ordinarily separate one species from another. And if this is true, if animals really can benefit emotionally and physically from the company of people, then it logically follows that humans should also be able to derive similar benefits from the company of animals. In which case, the evidence described in the previous chapter would support the conclusion that pet-ownership, like marriage, friendship or the parent–child relationship, can indeed have a measurable impact on human susceptibility to physical and mental disease.

Of course, relationships between people undoubtedly involve elements which are lacking or greatly attenuated in relationships between owners and their pets, and these elements may be crucial to our happiness and well-being. For one thing, humans are highly cooperative and, up to a point, altruistic. They help each other out in times of trouble in ways beyond the scope of most pets. The help we derive from other people takes a multitude of forms from practical assistance and helpful advice to financial and moral support, and most of us would agree that it is immensely reassuring to know that we can rely on such life-lines, if the worst comes to the worst. People also provide us with opportunities to unburden ourselves, and to communicate and share our worries, fears and frustrations. Children often use pets in this confessorial role, but as adults most of us need understanding friends, relatives or professional counsellors for the same purpose. Our very sanity may depend on our ability to communicate with others. Many psychologists view the human personality as an essentially fragile structure which, in order to remain whole and coherent, needs regular support and affirmation. Without the opportunity to compare our experiences with others or the chance to share and exchange opinions, attitudes and beliefs we can easily lose our bearings and become utterly disorientated.[43] Since pets cannot speak; since they cannot understand our opinions or advise us when we are on the right or the wrong track, it is difficult to see what they can contribute to this aspect of our psychological welfare. Yet the very fact that non-verbal animals can experience the same benefits from social affiliation as humans strongly suggests that the therapeutic effects of social support are to some degree independent of language.

Whenever one person interacts socially with another, they exchange a continuous stream of non-verbal messages; subtle nuances of posture, gesture and facial expression which convey all kinds of information both about themselves and about how each views the other. Skilled actors are able to reproduce these signals at will, but most of us transmit and receive non-verbal messages unconsciously. Perhaps because they are largely unconscious and therefore difficult to disguise, non-verbal signals may often convey a more honest and reliable impression of a person's true feelings than words. And this is particularly true of our feelings for others. We can tell someone that we value them, for example, or that we care about them, but the confession will sound hollow and untrustworthy unless it is backed up by appropriate non-verbal gestures.[44] Generally speaking, the people we like most in this world, and the ones whose company we find most rewarding and fulfilling, are the ones who give us the non-verbal impression that they like us in return.

The difficulty human beings have in concealing their true feelings is largely a good thing. Without it we would never be able to trust anybody. But it can also give rise to misunderstandings and a great deal of unhappiness. Chronically lonely people, for instance, tend to have negative feelings about themselves; they may see themselves as unattractive or as social failures, or as being unloved, unwanted, worthless or rejected. These inner feelings inevitably leak out non-verbally during social encounters, making such people appear either overdemanding and intrusive or, conversely, withdrawn and unfriendly. Either way, such signals can be extremely off-putting to other people. This is apparent from popular stereotypes of lonely people which are generally harsh and uncompromising. They are often perceived as intentionally reclusive, complaining, self-absorbed and self-pitying; as 'pathetic without being tragic', and as bringing their condition on themselves by being socially clumsy or inept. The lonely person, in other words, sends signals which make him appear less likeable, and he receives signals from others which merely confirm his worst opinions about himself. The net effect of all this is to exaggerate his sense of alienation and rejection still further.[45]

The findings of Boris Levinson, the Corsons, and others who have engaged in pet-therapy, suggest that it is precisely these kinds of isolated and socially unresponsive individuals who stand to benefit most from the companionship of animals. And this in turn suggests that companion animals somehow have the capacity to reconnect such people with the outside world by breaking down the vicious circle of non-verbal misunderstanding that surrounds them. This ability has a practical application in therapy, but it is also of wider concern since it implies that pets may be providing their owners with a special kind of emotional support which is lacking or at least uncommon in relationships between people.[46] How pets manage to perform this remarkable service can be readily understood when one examines in detail the way in which animals, such as dogs and cats, interact and communicate with their owners.

CHAPTER 8

Four-legged friends

Animals make such agreeable friends – they ask no
questions, they pass no criticisms.

George Eliot, *Scenes of Clerical Life*

The human species shares this planet with a dazzling variety
of different animals but it has only selected a tiny minority to
occupy the privileged position of pets. And among these
so-called 'pets' only a handful has been used for companionship. Rare, expensive and unusual animals are often employed
to advertise status. The wealthy business tycoon who keeps
a leopard as a pet may, conceivably, do so out of a genuine
affection for leopards. But, more likely, the animal is simply
intended as a signal of his power, affluence and ability to
flaunt convention. He is, in essence, a modern counterpart of
the earlier monarchs and aristocrats who used exotic animals
and menageries as emblems of status. Other 'pets' are kept
chiefly for their decorative potential. Brightly coloured
tropical fish and cage-birds can perform the same function
as kinetic art or mobile sculpture by adding colour and life to
otherwise drab and uninviting surroundings. Still other kinds
of 'pets' attract the specialist hobbyists who collect animals
like other people collect stamps. A few species can even
perform all of these different roles at the same time.
Coral-reef fish, for example, are simultaneously appealing to
the status-seeker, the interior decorator and the collector by
being both expensive, aesthetically pleasing and available in a
wide choice of varieties. It is, of course, possible to develop a
special affection for one particular individual coral-reef fish

or, indeed, any animal but, by and large, the person who is primarily seeking companionship will avoid strange and unconventional creatures and plump, instead, for some thoroughly ordinary, mundane, domestic species such as a dog or a cat. Dogs and cats vastly outnumber all the other kinds of companion animals in our society, and this leads one to wonder why on earth, out of all the species available to us, we choose to lavish so much affection on two medium-sized carnivores rather than on, say, squirrels, pigs or hyraxes.

In part, this choice is probably an accident of history; a consequence of the fact that dogs and cats, or wolves and wildcats, happened to be in the right place at the right time when our ancestors were taming and domesticating various wild mammals. They have also been a part of human society for thousands of years and have therefore had plenty of time to adapt to the role of companions. But neither of these factors explains why dogs and cats were favoured in the first place, or why these two species have managed to stay the course when most of the other mammalian pets tamed by our forebears have either fallen by the wayside or been converted into domestic meat-producers. The most likely explanation for this enduring success is that, right from the outset, wolves and wildcats possessed particular qualities which made them especially suitable and desirable as animal companions.[1]

Some of these characteristics are fairly obvious. For example, unlike the vast majority of domestic animals, dogs and cats do not need to be fenced in, caged or tethered in order to remain in the vicinity of their owners. Dogs develop specific attachments for particular individuals or groups of individuals, and cats display a powerful affinity for particular places and, to a lesser degree, for people. They share these characteristics in common with their wild ancestors. Wolves develop strong social bonds with each other, and wildcats have a tendency to remain strictly within their own home-ranges or territories.[2] Perhaps because of their habit of remaining most of the time within one specific area, wolves and wildcats are also relatively hygienic; generally depositing their urine and faeces in particular places, often on the boundaries of their territories and away from the central den area. Fortunately for us and for them, their domestic descendants retain this habit. We may, understandably, object to canine and feline excrement in our gardens and parks, but at least we do not normally have to contend with this problem inside our homes or on the carpet. The situation would be very different if we gave, say, monkeys, birds or guinea-pigs the run of our houses. These animals are notoriously difficult to house-train for the simple reason that in the wild they drop their waste products wherever they happen to be. One reason why we tend to keep animals of this kind caged or tethered is that it gives us some control over the mess they produce.[3]

Dogs and cats are also active during the day, especially around dawn and dusk, when people also tend to be up and about. Anyone who has ever tried sleeping in the same room as a hamster will recognize the advantages of this diurnal habit, and it also means that we are ready to interact with these pets at about the same time that they are ready to interact with us. It is surely significant that nocturnal rodents such as mice, rats and hamsters have only become popular as pets since the advent of the electric light; an invention which has artificially extended the period of human activity into the hours of darkness when these animals are normally active.[4]

Even the physical size of dogs and cats is probably important to their success as companions. They are large enough for us to view them as recognizable individuals and treat them, so to speak, as small people. But they are also small enough so that the majority do not pose a serious threat. By nature both species are predatory carnivores and they retain the ability to cause us physical harm should they choose to do so. With cats the danger to life and limb might seem fairly negligible but with dogs, especially large dogs, it is quite a different story. Dominance-related aggression – attempts by dogs to become socially dominant to their owners – is one of the commonest behaviour problems among pet dogs, and it often results from the owner failing to assert himself in situations involving some sort of conflict of interests.[5] Clearly, if good relations are to be maintained between people and dogs, it is essential that we retain the upper hand. And one thing which helps us to do this is our superior size and strength. The other thing which helps is the dog's natural inclination to defer to individuals whom he perceives as dominant. This tendency to accept a subordinate rank is presumably something which dogs have inherited from the days when they lived in hierarchically organized packs.[6] The dog's characteristic eagerness to please and willingness to cooperate is another popular legacy of pack life. Unruly, disobedient dogs can be a liability and a menace, and it is most unlikely that humans would have tolerated dogs for as long as they have if this species was deficient in intelligence and trainability.

Characteristics like these have enabled dogs and cats to fit fairly unobtrusively into the structure of human society without drastically upsetting our routines or disrupting our way of life. Yet, although these attributes must, ultimately, be important in maintaining good relations and reducing conflict between people and their pets, they are also, in a curious way, peripheral.

To an outsider, many pets would seem to have few redeeming features. They are too large and obstreperous, they urinate on the carpets, run away, wake the household or the neighbours, are stupid and untrainable, and absurdly greedy or fussy over their food. Yet their owners seem to have an almost unlimited capacity to forgive and forget; to go on loving them despite everything. This strongly suggests that animals such as dogs and cats and, to a lesser extent, many other companion animals are exploiting some other, more subtle channel into our affections. Human relationships are as complex and variable in quality as the individuals who take part in them, and it is difficult to generalize about what distinguishes good and satisfying relationships from bad ones.[7] Nevertheless, a number of fundamental pre-conditions need to be met in any relationship before a mutually rewarding and fulfilling dialogue can be established between two individuals. Prominent among these are: mutual liking – the belief that the other person likes, respects and values one,

preferably as much as one likes, respects and values them –
and mutual trust – the knowledge that the other person can
be relied upon not to take unreasonable advantage of one's
affections.[8] Both of these ingredients can be transmitted
verbally, simply by telling someone that we like them or that
they can trust us but unfortunately, as every politician knows,
the language of words is notoriously deceptive and unreliable.
For this reason humans, like other animals, depend to a major
extent on non-verbal signals when deciding whether or not
another individual is worthy of trust and companionship.

Some years ago, I was interested in discovering what
particular aspects of dog behaviour were important to the
majority of dog-owners. I distributed a simple questionnaire to
an assortment of Cambridge dog-owners in which they were
asked to rate their own pets, together with some hypothetical
'ideal' pet, in terms of twenty-two different aspects of canine
behaviour; aspects such as playfulness, obedience, separation
distress, and affection which I knew to be of interest to most
owners. In brief, the following results emerged: the hypothetical 'ideal' dog proved, not unexpectedly, to be a paragon
of canine virtue. It loved going for walks, was superbly
obedient and intelligent, and was exceptionally affectionate,
welcoming, expressive and visually attentive to its owner. It
was also moderately playful, territorial, protective, friendly to
strangers and sensitive to its owner's moods. It was not too
nervous, excitable, possessive or greedy, and it rarely showed
signs of extreme separation distress. More interesting,
however, were the handful of attributes for which real or
actual dogs closely matched their owners' ideal expectations.
Prominent among these were the tendency to be very
affectionate, to welcome the owner intensely whenever he or
she came home, to be highly expressive (almost human), and
to attend closely to everything the owner said or did. In other
words, out of the twenty-two possible attributes, these owners
unconsciously attached greatest importance to the four which
were concerned with non-verbal messages of love and attachment.[9] Dogs and cats are particularly adept at conveying these
apparent signals of friendship, and their ability to win friends

and influence people owes more to their skill in this direction than to any other factor.

Perhaps the most obvious way in which these animals signal their liking for us is by their habit of seeking out our company and remaining near or even in physical contact with us. The average dog, for example, behaves as if it is literally 'attached' to its owner by an invisible cord. Given the opportunity, it will follow him everywhere, sit or lie down beside him or at least close by, and exhibit clear signs of distress if the owner goes out and leaves it behind, or shuts it out of the room unexpectedly. In some dogs this kind of separation distress can become exaggerated, even pathological. One of the more common behaviour problems reported by dog-owners is the animal which barks, whines, defecates, urinates or chews up and destroys furniture and household fittings whenever it is left alone for any length of time. Not surprisingly, these symptoms are more typical of young dogs and puppies, but they may persist into adulthood, particularly in animals which are unused to being left on their own, or in those whose early attachments have been disrupted through abandonment or re-adoption.[10] In such cases, the resemblance to emotionally insecure children from disturbed family backgrounds is too obvious to be ignored.

Separation anxiety is much less of a problem with cats, although it undoubtedly occurs especially in young cats and kittens. On the whole, as befits their less social origins, cats show greater independence than dogs and, when permitted, will spend much of their time patrolling the neighbourhood on a solitary basis. Yet they are quite capable of recognizing their owners at least at close range, and will actively seek proximity and physical contact with people when they are not actually asleep or out on the prowl.

When a human being is in love with or strongly attached to another individual he may tend to feel possessive and jealous when he perceives that the relationship is threatened by the attentions of a third party. In other words, signs of jealousy are also signs of love. We are inclined to think of jealousy as a strictly human emotion, although pet dogs and cats exhibit behavioural symptoms which, in every respect, are remarkably similar. A study carried out by psychologists at the University of Western Illinois explored this phenomenon, and found that 79 per cent of cat-owners and 91 per cent of dog-owners reported signs of jealousy in their pets. By far the most common cause of jealous and possessive behaviour for both species was the owner giving attention to a third party, such as another animal, a person or a stuffed toy animal. The most common symptoms were attention seeking, aggression directed at the third party, withdrawing and 'sulking', hyperactivity, and even apparent acts of revenge (in the case of three dogs) against the owner. The tendency of pets to display jealousy was also positively associated with how attached they were to the owner.[11] Extreme separation anxiety and possessiveness can create problems between owners and their pets, just as they can in relationships between people. In rare cases, canine jealousy has even resulted in serious or fatal attacks, particularly on infants and young children.[12] Nevertheless, both these behaviour patterns signal the pets' dependence on the owner and, in moderation, people undoubtedly find this sense of being needed rewarding.

In addition to wanting merely to be near their owners, cats and dogs also derive considerable pleasure from being stroked, petted and caressed. If ignored for too long, many dogs will actively solicit this kind of physical attention from people by pawing them, jumping up on them, nuzzling their hands or laying their heads emphatically on someone's lap or knee. Some dogs have the endearing habit of rolling over on their backs, supposedly 'to have their tummies rubbed'. The truth is perhaps less appealing. In reality, rolling over, in this context, is probably derived from the posture puppies adopt when allowing the mother to clean their anogenital regions with her tongue. In adult dogs and wolves, the behaviour constitutes an act of passive submission to a more dominant individual. In extremely submissive individuals it may be accompanied by urination, as in puppies.[13]

Cats solicit physical contact by rubbing themselves against people or other cats; sometimes with the whole body, but more especially with the side of the face and head. The origin of this behaviour is obscure, although the area around a cat's face and mouth is studded with tiny glands which may be depositing the animal's scent on the owner as it rubs. Whatever the function, rubbing seems to be a tremendously stimulating activity for cats, and some individuals engage in an ecstasy of rubbing when scratched around the head and throat. Cats generally display their pleasure in physical contact by purring; a mysterious and curiously appealing guttural rumble produced deep down in the throat. Many cats also 'knead', unsheathing their claws slightly and treading rhythmically up and down with the front paws. A few cats also salivate and dribble slightly during this ritual and some will actually nuzzle or nibble the owner at the same time. Kneading is particularly stimulated by soft substrates such as thick carpets or woollen clothes, and this, together with the salivation and nuzzling, provides a clue to the origin of the behaviour. Soon after birth, kittens begin purring and kneading the soft fur of the mother's belly during suckling. In this situation, the kneading may serve to stimulate the flow of milk to the mother's teats. In other words, affectionate physical contact with the owner seems to reactivate many of the cat's infantile suckling reflexes.[14]

Like small children, cats and dogs are uninhibited in their demands for physical attention and comfort, and pet-owners, like parents, respond to these spontaneous demands by relaxing and becoming less physically inhibited themselves. Living as we do in a society where physical intimacy between adults is strongly repressed, at least in public, pets provide us with a valuable and socially acceptable outlet for the expression of intimate behaviour.[15]

Greeting behaviour is another area in which dogs and, to a lesser extent, cats excel. Greetings and reunions are highly conventional and stereotyped in the human species, and they form an essential starting point for most social interactions. The raised eyebrows of recognition, the approach, the handshake or embrace, and the various verbal utterances all combine to signal a reaffirmation of existing social bonds. And, generally speaking, the quality and intensity with which these signals are expressed provides a fair measure of the depth of feeling that exists between the 'greeter' and the 'greetee'.[16] I once heard someone say that the reason he liked his dogs so much was that whenever he went out to the shops for five minutes, they invariably greeted him as if he had been away for five years. It is easy to see how dogs give their owners this impression. The typical canine welcome begins the instant the animal hears or sees the owner approaching. Immediately the tail starts wagging, the head and ears are lowered, the eyes are closed slightly, and the sides of the mouth are drawn back into the so-called 'submissive' grin. The animal rushes toward and around the owner with the tail, by now, wagging so enthusiastically that the animal's entire hindquarters become incorporated into the movement. Many dogs produce small, breathless yelps of excitement at this stage and, unless they are very well disciplined, most will attempt to jump up and lick the owner's face. In fact, it is difficult to convey in words the extraordinary intensity and emotional exuberance of these performances.

Greeting ceremonies in wolves closely resemble those of dogs and, once again, the sequence of actions appears to be derived from elements of pup behaviour. During weaning, wolfpups are gradually deterred from attempting to suckle and, instead, are encouraged to solicit the regurgitation of solid food from parents and other members of the pack. To do this they crowd excitedly round the adult's head and jump up and lick the corners of its mouth. The performance is also accompanied by a lot of tail-wagging and various gestures of submission. In adult wolves, these same sequences of behaviour develop into ritualized displays of mutual greeting and pack solidarity. Most of the attention is directed toward the highest ranking individuals who, obligingly, act out the parental role by parading around with bones, sticks and other inedible objects in their mouths.[17] Thus, when a dog jumps up in an attempt to lick the owner's face – a gesture which is often interpreted as a sort of canine kiss – it is actually re-enacting the infantile food-begging ritual.

Feline greetings, though generally less histrionic, are

nevertheless unmistakable. The initial signal of recognition, often in response to a vocal greeting from the owner, is generally a soft, bird-like chirp; a sort of cross between a purr and a miaow. As the cat approaches or is approached, the tail is abruptly elevated and held vertical, sometimes with end curled over into a question mark. The cat then usually rubs itself against the owner's hand or legs, purrs loudly and, occasionally, rolls around on its back. Again, all these signals also occur in greetings between kittens and their mothers and, occasionally, between adult cats.[18]

Clearly, these signals transmitted by dogs and cats are not the same or even, necessarily, similar to those given by people in comparable situations. Humans do not possess tails, or mobile ears; nor do we roll around on our backs, or purr during social interaction. But this is largely immaterial. Differences in communication patterns between people and cats and dogs inevitably lead to the occasional misinterpretation, especially by persons who are unaccustomed to interacting with these animals, or by animals that are unaccustomed to dealing with people. Even the gentle kneading of a cat's claws can be disconcerting to someone who is unfamiliar with cats. But this sort of failure to understand is essentially no different from the misunderstandings which would arise between, say, a Frenchman and an Englishman. As with words, the form or structure of many non-verbal signals conveys little meaning in itself. The meaning of a signal is determined not primarily by its sound or appearance, but by the context in which it is used and the intensity with which it is expressed. For this reason, a person can learn to decipher any language, even animal language, as long as the appropriate contextual and intensity cues are available for interpretation. Humans greet each other, they seek proximity with individuals they like and, in the right setting, enjoy physical contact and intimacy. The fact that dogs and cats do these things in qualitatively different ways is irrelevant. What is important is that they do these things in ways which are obvious, unambiguous and readily interpreted. It does not matter that a dog may be inviting its owner to regurgitate food when its licks his face. What matters is that the behaviour occurs in the context of a reunion, and that the owner interprets this as an expression of joy at his return. Likewise, it does not matter to the dog when its owner fails to regurgitate food or parade around with a bone in his mouth. Although the stereo-typed greeting gestures of dogs are probably influenced by genetic factors, their ability to understand signals, even those emanating from another dog, is acquired primarily through the experience of interacting and communicating with others. A wolfpup that is brought up in isolation is unable to under-stand the gestures of normal wolves. But it can acquire the necessary understanding within a few days of social contact.[19]

Although human non-verbal signals, such as hand-waving, kissing and embracing, sometimes involve overt movements and actions, the majority are conveyed by subtle and transitory changes in facial expression. The eyes, in particular, provide one of the most important channels for communicating emotion and affect. Humans are virtually unique among primates in using eye-to-eye contact as a means of expressing intimacy, as well as aggression and dominance.[20] Normal adult rhesus monkeys, for example, almost always interpret mutual eye contact as a gesture of hostility, and they do everything they can to avoid it. Subordinate monkeys in a social group glance nervously and surreptitiously at dominant individuals, and look away immediately if their glances are reciprocated. Conversely, dominant monkeys often assert themselves by staring rivals down; sometimes with a simple look, sometimes with an exaggerated scowl.[21] Comparable patterns of gaze and gaze avoidance are found in humans. When we admonish children for staring at strangers, it is because we are aware that being stared at can be upsetting and annoying, even when the stare is motivated by innocent curiosity. The stare is disconcerting because looks involving that kind of intensity and duration often denote hostility. When we are furious we tend to look intently at the objects of our rage. Looking away from someone signals a 'cut off' – a desire to avoid confron-tation. We also avoid looking at people from whom we anticipate anger or disapproval. This accounts for the guilty or 'shifty' looks that people tend to acquire when they are, say, perpetrating some sort of deception. Patterns of gaze are also used to express dominance. In hierarchically organized social groups of people, low-ranking individuals tend to look more at superiors than vice versa, and individuals who are looked at most see themselves and are seen by others as the most powerful members of the group. In face-to-face encounters, subordinates look more at dominants, but they are more likely to break mutual eye contact first and more frequently. Dominants look less, but their looks are more prolonged and they are less likely to be the first to break off eye-to-eye contact.

Yet, despite all its uses in threat and dominance interactions, looking in the human species is also a signal of liking. Numerous studies have shown that people look more at those they like and, in the appropriate circumstances, being looked at is perceived as a signal of interest and attraction. The phenomenon is obvious between lovers, but it is also of enormous importance in the majority of social relationships, although most of the time we are unconscious of the fact. Social psychology experiments, in which subjects were interviewed in pairs, have shown that the subjects who were looked at most by the interviewer believed that they were preferred. Other studies have demonstrated that when one individual looks a lot at another, not only does the other see this as a signal of liking, but he also tends to like the 'looker' more in return.[22] Many of these findings, of course, are already enshrined in the English language in phrases such as 'they only had eyes for each other', and 'having the highest regard for', or in words like 'admire' (L. *ad*, at; *mirare*, to wonder) and 'respect' (L. *re-*, back; *specere* to look).

The role of eye contact in affiliative behaviour becomes established in early infancy during face-to-face interactions between mother and child. Human infants show what seems to be a strong attraction for faces or pictures of faces, and when given the choice between two television images of a person's face, one in which eye-to-eye contact is possible and the other impossible, they strongly prefer looking at the former. Mothers also behave in ways which appear designed to achieve mutual visual contact. They tend to move themselves or the infant so as to remain within the child's visual field, they maintain an optimal distance from the infant's face to allow for his limited focusing abilities, and they make sounds and gestures to attract the baby's attention. When mutual eye contact is established, the mother will 'reward' the child with greeting signals, such as eyebrow raising, smiling and high-pitched vocalizations.[23] Studies of blind infants who are unable to maintain eye contact with their mothers have shown how important this visual dialogue is to the formation of a close mother–child relationship. To quote from one author on the subject:

Blind eyes that do not engage our eyes, that do not regard our faces, have an effect upon the observer which is never fully overcome. When the eyes do not meet ours in acknowledgement of our presence, it feels curiously like a rebuff. Certainly mothers attribute 'knowing' and 'recognition' to a baby's sustained regard of the face long before he can actually discriminate faces, and this is only because the engagement of the eyes is part of the universal code of the human fraternity, which is read as a greeting and an acknowledgement of 'the other' long before it can have any meaning for the infant.[24]

The frequency and patterning of gaze and mutual gaze also play a crucial role in relationships between people and their pets. Although they do it to different degrees and in different ways, both dogs and cats regularly engage in mutual eye contact with their owners, and spend considerable amounts of time observing their activities. Owners also deliberately solicit eye-to-eye contact with their pets by attracting their gaze, usually by calling the animal's name, and then sustaining their attention by talking to them. Cats sometimes exhibit a particularly frank and detached sort of stare which, perhaps because it is lacking in deference, some people find disconcerting. Nevertheless, when a cat is stared at directly, particularly at close range and without any other kind of interaction, it will generally close or half-close its eyes a few times before slowly looking away. This visual cut-off is easily observed in encounters between adult cats, and it denotes a mild form of submission or at least a desire to avoid conflict. To most cat-owners, however, the signal has a different meaning. The half-closed eyes, combined with the animal's round face, and the turned up corners of its mouth produce a pleasing facsimile of a rather plump and self-satisfied human smile.

A cat sitting on a lap while being stroked and caressed will often stare up into the owner's face with a wistful or adoring look. The effect is also greatly enhanced by the animal's large and often beautiful, almond-shaped eyes. In humans and many other animals the pupils of the eye dilate when looking at something attractive or appealing. There is abundant scientific evidence that this pupillary dilation reflex acts as a signal of interpersonal attraction and, at one time, ladies of fashion used to enlarge their pupils artificially with belladonna (meaning 'beautiful lady') to make themselves more attractive.[25] The vertical pupils of cats are able to open much wider than those of a human and, although I know of no evidence that cats' pupils enlarge when they look adoringly at their owners, I strongly suspect that this is the case.

Dogs, in general, look at their owners much more than cats do, and in a different way. Jack London, perhaps one of the keenest literary observers of canine behaviour, describes the situation exactly in the following account of the dog Buck in *The Call of the Wild*:

He would lie by the hour, eager, alert, at Thornton's feet, looking up into this his face, dwelling upon it, studying it, following with keenest interest each fleeting expression, every movement or change of feature. Or, as chance might have it, he would lie farther away, to the side or rear, watching the outlines of the man and the occasional movements of his body. And often, such was the communion in which they lived, the strength of Buck's gaze would draw John Thornton's head around, and he would return the gaze, without speech, his heart shining out of his eyes as Buck's heart shone out.[26]

A dog that is stared at directly by its owner will react quite differently from a cat. At frequent intervals it will slightly avert its gaze, as if it finds the experience of mutual eye contact at close quarters unnerving. Ordinarily, this manner-ism produces an endearing quality of modesty, like an oriental courtesan. But when exaggerated – for example, when the dog is expecting punishment for some misdeed – this kind of gaze avoidance can look decidedly guilty. Wolves also communicate by means of eye contact and, as in rhesus monkeys, a direct look is generally construed as an aggressive challenge and is avoided by subordinate individuals. In wolves, the iris of the eye is yellow in contrast to the black pupil and this, presumably, makes it easier for them to detect gaze direction and gaze avoidance in others.[27] It is probably significant, in this regard, that most domestic dogs have been bred selectively to have dark eyes and irises. Pale or yellow-eyed dogs tend to be disliked by dog-breeders because they have a reputation for being sly and untrustworthy. One wonders how much of this reputation is due to the fact that these animals' normal guilty or 'shifty' looks are simply more conspicuous. The cat's eyeball is incapable of the degree of lateral and up-and-down movement found in dog or human eyes. Hence, a cat that wishes to avoid direct eye contact must either close its eyes or turn its head away.

Looks by themselves, of course, cannot be used to determine a person's attitudes or intentions. A look signals attention and interest, but in the absence of other cues, we have no way of judging whether someone is interested in embracing us or stabbing us. To obtain this kind of information we need to absorb and interpret a wealth of associated messages, such as our relationship with the individual who is doing the looking, his overall demeanour, and the facial expression which accompanies the look.[28] All of this is equally true of looks we receive from cats and dogs.

Although they can alter the angle of their ears and whiskers, and open and close their eyes, cats have relatively blank and immobile faces. This is no reflection on the character of the cat; it simply means that cats do not possess the musculature needed to produce a highly expressive face. Despite this short-coming, cats' faces, except when they are angry or alarmed, seem to be permanently fixed in an expression of slightly detached contentment which is intrinsically appealing, at least to cat-lovers. Dogs' faces, in contrast, are richly endowed with facial musculature, and they use these muscles to express emotion in ways similar, or at least comparable, to the way humans use them.[29] Dog-owners are highly attuned to their pet's facial expressions and, rightly or wrongly, interpret these expressions as if they are coming from another person. Hence people readily talk about dogs looking happy, sad, bored, reproachful, disdainful, loyal, honest, kind, devoted, surprised, embarrassed and mischievous, using precisely the same adjectives they would use to describe human looks. Certain canine expressions have been enhanced, probably unconsciously, by artificial selection. The wrinkles and folds of skin around the eyes of bloodhounds and boxers, and the drooping jowls of basset hounds and spaniels produce permanently forlorn and reproachful expressions which appear specially designed to evoke human sympathy and indulgence.

Even the angle of eye contact we exchange with dogs and cats conveys meaning. These animals do not just look at us: they look up to us, and all that this implies. Their smaller stature and necessarily subordinate status relative to the owner, automatically places them in the junior position. If John Thornton's dog, Buck, had been human, his behaviour would have constituted an act of submissive adoration; of hero-worship. His looks were the looks of a younger, lower-ranking individual toward a much loved and admired senior. Aldous Huxley once wrote that 'to his dog, every man is Napoleon',[30] and it is undoubtedly true that many owners, consciously or otherwise, derive a certain satisfaction from this kind of VIP treatment. It is easy to be cynical and dismiss pet-owners as people who use animals to boost their own fragile egos. But one must also bear in mind that everybody likes to be thought highly of and there is nothing intrinsically wrong with this unless, as occasionally happens, the privilege is abused.

Finally, and perhaps most importantly, dogs and cats are speechless. Most people talk to their pets but, although they may listen attentively and even respond appropriately to certain words and phrases, they cannot talk back. We cannot converse with them about the political situation in Uruguay, we can't discuss the weather, or ask their advice on what to wear. We cannot share with them our beliefs, our aspirations or ideas. They are forever mute and unresponsive to the vast intellectual and linguistic component of the human condition. Yet, far from being a drawback, this may be one of their most endearing assets as companions. Most pet-owners believe that their animals are sensitive to their moods and feelings, and many confide in their pets verbally. In other words, the animal is perceived as empathic. It listens and seems to understand, but it does not question or evaluate. Aaron Katcher, an American psychiatrist, has compared this essentially one-way dialogue between owner and pet with the kind of empathic, non-interventional relationship which certain types of psychotherapist attempt to establish with their patients. Furthermore, pets have an added advantage over human therapists in that one can stroke and cuddle them at the same time.[31]

The ability to transmit complex information using written and spoken language singles the human species out from other animals. It is one of the foundation stones on which the entire edifice of human culture and civilization rests. Unfortunately, it is also regularly used for less creative purposes. Language endows humanity with the power to convey unreliable as well as reliable information. It enables us to distort the truth, to prevaricate and to deceive and flatter our fellows to a degree unprecedented in the animal king- dom.[32] Language is also the weapon of criticism. In this guise it undoubtedly has a constructive role to play but, all too often, it is merely used to undermine and destroy. Lacking the power of speech, animals cannot participate in conversation or debate but, by the same token, they do not judge us, criticize us, lie to us or betray our trust. Because they are mute and non-judgemental, their affection is seen as sincere, innocent, and without pretence. They are essentially reliable and trust- worthy.

To summarize, then, dogs and cats have maintained their

popularity as animal companions not, primarily, because they are home-loving, active during the day, non-aggressive or easy to house-train. These things are important but they are not, in any ordinary sense of the word, companionable. The indomitable success of these two species is chiefly owing to their powers of non-verbal expression. By seeking to be near us and soliciting our caresses, by their exuberant greetings and pain on separation, by their possessiveness and their deferential looks of admiration, these animals persuade us that they love us and regard us highly, despite all our manifest deficiencies and failures. However much we may regret the fact on occasions, we humans need to feel liked, respected, admired. We enjoy the sensation of being valued and needed by others. These feelings are not trivial, nor are they a sign of weakness. Our confidence, our self-esteem, our ability to cope with the stresses of life and, ultimately, our physical health depend on this sense of belonging. Without it, existence would be hollow and without purpose. Of course, all other things being equal, a positive, affectionate relationship with another person is more rewarding and satisfying than a relationship with a cat or a dog or, indeed, any other kind of pet. It has more depth, more scope, more complexity, and it is altogether more enriching. But inter-personal relationships can be a source of pain as well as pleasure. We are potentially in conflict and competition with other people in ways that we do not conflict or compete with our pets. Humans are consummate actors, capable of concealing and disguising their true feelings and intentions when it suits their purposes to do so. For this reason we exercise extreme caution when choosing our friends for, otherwise, we would lay ourselves open to the possibility of deceit, manipulation, betrayal and rejection, and all the hazardous medical consequences that arise from these kinds of negative interaction.[33] Our relations with companion animals may be shallow by comparison, but at least their affection for us is reliable and unconditional, and it is patently absurd to argue that this kind of constant emotional support is either inconsequential or unhealthy.

The signals of affection and attachment conveyed by pets

are, undeniably, very similar to those displayed by infants and young children toward their parents. There is also little doubt that humans have exaggerated or enhanced the neotenous, child-like qualities of companion animals through generations of unconscious selection. But, again, these are not sufficient reasons to describe pets in derogatory terms as child substitutes or social parasites, or to label the human–pet relationship as abnormal or perverted. The American social psychologist, Robert Weiss, has described 'the opportunity for nurturance' as one of the most important and fulfilling aspects of many human relationships.[34] Taking on the responsibility for the care and well-being of a child, for example, can help people to develop a sense of being needed: it can provide meaning to their lives, and help them to sustain commitment to personal goals. By virtue of their resemblance to children, pets can undoubtedly provide their owners with comparable psychological rewards.

Pets do not just substitute for human relationships. They complement and augment them. They add a new and unique dimension to human social life and, thereby, help to buffer their owners from the potentially numbing and debilitating effects of loneliness and social isolation. Perhaps, in the best of all possible worlds, it would be preferable if humans satisfied all their social and affiliative needs with each other. The world would no doubt be a happier and saner place as a result. But, until that Utopian day dawns, we can do a great deal worse than seek the partial fulfilment of these needs in the company of animals.

PART IV

Exploitation and sympathy: a conflict of interests

CHAPTER 9

The myth of human supremacy

Animals, whom we have made our slaves, we do not like
to consider our equal.

Charles Darwin, *Letters*

So far this book has been largely devoted to a discussion of
popular beliefs, theories and misconceptions about pet-
keeping. In one way or another, most of these beliefs tend
to denigrate the human practice of keeping animals for
companionship by implying either that pets are merely
substitutes for people; that pet-keeping is an unnecessary and,
therefore, wasteful activity, or that it is a sort of pathological
condition arising from the human tendency to respond in a
nurturant, parental manner to young or dependent-seeming
animals. All of these ideas stem from the supposition that
pet-keeping is essentially 'useless', that it serves no practical
purpose – and one of the main effects of this pragmatic or
economic bias has been to discourage research into the nature
and origins of this widespread, and in many ways puzzling,
phenomenon.

As I hope I have demonstrated, none of these prejudicial
notions about pet-keeping is borne out by the available
evidence, and for this reason, an alternative view has been
proposed: that far from being perverted, extravagant, or the
victims of misplaced parental instincts, the majority of
pet-owners are normal rational people who make use of animals
to augment their existing social relationships, and so enhance
their own psychological and physical welfare. Thought of in
these terms, keeping a dog, say, for companionship is no more

outlandish or profligate than wearing an overcoat to keep out the cold. Indeed, it would be fair to argue that pet-keeping is genuinely 'adaptive' in the evolutionary sense of the word, since it contributes to individual health and survival by ameliorating the stresses and strains of everyday life. Pets, according to this view, are as useful in their way as domestic pigs or poultry. The only difference being that one cannot easily measure their usefulness using a conventional economic yard-stick, any more than one can attach a precise value to friendship.

Unfortunately, this alternative theory also has its weaknesses. If keeping animals such as dogs and cats for companionship is truly beneficial, why then is not the practice even more widespread than it is? Pet-keeping has developed unprecedented levels of expression in the industrial West, but it is poorly developed or virtually non-existent in many other societies around the world. Similarly, while roughly half the households in Europe and North America own a pet, an approximately equal number do not; a somewhat unexpected state of affairs given the proposed benefits of pet-ownership. All of these facts suggest that either the theory is wrong, or that other forces are at work which limit or control the spread of the pet-keeping habit.

Some of these forces have already been discussed in previous chapters. Inadequate housing, insufficient time to care for an animal and economic constraints can all provide reasons for not keeping a pet, and it is clear that many would-be pet-owners are dissuaded from doing so by these kinds of circumstances.[1] But this still leaves a substantial residue of people who are quite capable of keeping pets but who seem, nevertheless, to be indifferent or opposed to the postulated delights of animal companionship. Individual experience probably contributes to this sort of antipathy. The results of a study outlined earlier in this book suggested that pets tend to run in families. Children brought up in homes containing pets are much more likely, on average, to become pet-owners as adults than those who did not have this sort of early experience. Moreover, the pro-pet attitudes created by early contact show marked specificity. People brought up with dogs tend to remain dog-lovers, those brought up with cats prefer cats, and those raised with both remain affectionately disposed toward both species.[2] It is not known whether this effect is due to some formative influence of the pet itself, or merely a product of different sorts of family environment, but either way the long-term outcome is the same: the practice of keeping animals as pets tends to pass down through families from parents to children, from one generation to the next. Conversely, people without the experience of associating with companion animals during childhood tend to remain relatively indifferent or resistant to their charms, thus introducing a sort of cultural inertia to the system. Such effects of early experience may help to explain why pet-keeping has not spread more rapidly and completely than it has, but it also raises other problematical questions. Why, for instance, has the practice spread only recently and not before? And why has it failed to spread in many other regions of the world? One possible answer to these questions is simply that some individuals and societies have no need to supplement their existing social relationships by making use of animals.

In attempts to explain the recent growth of pet populations in western countries, several writers have stressed the possible impact of social change.[3] Whereas, less than a hundred years ago, the majority of people lived out their lives within relatively small, stable communities, surrounded by several generations of close kin, the technological advances of the twentieth century have dramatically increased human mobility, and brought about the disruption or fragmentation of traditional family and community structures. This trend, it is argued, has increased the need for alternative sources of emotional support and companionship, particularly within the socially alienating and depersonalizing environments created by large cities. Significantly, the same arguments have been used to explain the increasing importance of friendship in modern social life. The psychologist Damon, for example, states that in societies where the family has been eroded by mobility, 'friendship replaces kinship in establishing relations between persons based on positive feeling, trust and other non-tangible manifestations of affection'.[4] This idea makes intuitive sense but, unfortunately, it is lacking in supporting evidence. Pet-keeping seems to be the norm rather than the exception among hunting and gathering societies (see chapter 4), yet hunter–gatherers live typically in small, stable and relatively closely knit kin groups in which, according to the theory, additional sources of companionship ought to be superfluous. Similarly, the results of various surveys and studies that have been conducted in the West suggest that *per capita* pet-ownership is actually lower among the more solitary and isolated members of society than it is among intact families with children.[5] Also, pet-keeping is no more frequent in urban than in rural areas although, according to some evidence, people who live alone or in cities *do* tend to form closer relationships with companion animals.[6]

Fortunately, an alternative explanation exists for all of these apparently anomalous findings which does not rely on arguments about immediate causation. It is entirely possible, for example, that the recent growth of the pet-keeping habit in western society is not so much a product of increasing need, but rather the inevitable outcome of historical changes in attitude, not only to pets, but to animals in general.

Western perceptions of and attitudes toward animals have their roots firmly embedded in the Judaeo-Christian philosophical tradition. According to this tradition, the Earth and the animal and plant species which inhabit it were created specifically to serve the interests of humanity.[7] The biblical account of creation provides implicit support for this view. In the first chapter of the Book of Genesis, God distinguishes humans from animals by creating the former 'in his own image'. At the same time He awards Man 'dominion over every living thing that moveth upon the earth'. Man lost some of his control over nature following the Fall, but was firmly reinstated after the Flood when God informed Noah that:

the fear of you and the dread of you shall be upon every beast of the earth, and upon every fowl of the air, upon all that moveth upon the earth, and upon all the fishes of the sea; into your hands are they delivered.

Thus equipped, the vast majority of Christian theologians have maintained that Man was the penultimate* expression of God's Divine Plan, and that all other species were inferior and subordinate to His will. Although the Bible endowed humans with privileged status compared with the rest of brute creation, it also repeatedly extolled the virtues of kindness toward animals – 'a righteous man regardeth the life of his beast' (Prov.12: 10) – and some early Christians evidently took these recommendations to heart. St John Chrysostom (AD 347–407), anticipating Darwin, argued that we should show great kindness and gentleness to animals 'for many reasons, but, above all, because they are of the same origin as ourselves'. St Francis of Assisi (?1181–1226) and some later members of the Franciscan Order also preached the love of nature and a relatively humane attitude to other species.[8] Unfortunately, during the Middle Ages this charitable outlook went drastically out of fashion.

In his recent book *Animal Minds & Human Morals*, the classical scholar, Richard Sorabji, attributes the relatively negative medieval view of (non-human) animals primarily to the influence of the Greek philosopher, Aristotle. Like his teacher, Plato, Aristotle exalted intellect or the power of reason above all other human faculties. Unlike Plato, however, he perceived the natural world as a sort of intellectual hierarchy – the so-called *Scala Naturae* or Ladder of Nature – with humans at the peak or top rung of the ladder, and animals and plants at various levels below this according to their reasoning abilities. Aristotle also argued that 'Nature does nothing in vain'. In other words, Nature is purposeful; it designs organisms for specific purposes, and the undoubted purpose of lower organisms was to serve as food or labour for those higher up the scale. Aristotle used the same argument to condone the Grecian slave trade. Some humans, he claimed, were 'natural slaves' in as much as they possessed lesser reasoning abilities than Greeks. Animals were even less capable of reason than human slaves, and they were therefore

* The ultimate expression was, of course, Jesus Christ.

deemed to be inferior and imperfect beings designed to serve the interests of more perfect and intelligent humans.[9] This emphasis on the irrationality of animals was subsequently adopted and modified by various Stoic, Epicurean and Neoplatonist philosophers, some of whom not only denied reason and belief to animals, but also the ability to hold concepts, and even the capacity to learn from past experience. Eventually, at the beginning of the fourth century AD, the same ideas were incorporated into early Christian tradition.[10]

In a discussion of the biblical commandment, 'Thou shalt not kill', St Augustine (AD 354–430) stated that people should not make the mistake of applying this rule to 'irrational living things, whether flying, swimming, walking or crawling, because they are not associated in a community with us by *reason*, since it is not given to them to have *reason* in common with us. Hence it is by a very just ordinance of the Creator that their life and death is subordinated to our use.'[11] According to Sorabji, the position adopted by Augustine, and subsequently by the majority of Christian authorities, was merely one of many on a question that had been hotly debated throughout the ancient world. The Pythagoreans and Platonists who preceded Aristotle believed, for example, not only that animals possessed rational souls, but that they were also the repositories of reincarnated human souls. Aristotle's own successor, Theophrastus (322–287 BC), rejected animal sacrifice and meat-eating on the grounds that animals were kindred beings, while Diogenes of Sinope (fourth century BC), the anarchic founder of a group known as the Cynics, made the radical claim that animals were in fact *superior* to humans in most respects. Writing only a century before Augustine, the Neoplatonist, Porphyry of Tyre (AD 232–309), brought all of this earlier material together in a remarkable synthesis entitled *On Abstinence from Animal Food* in which he argued that we owe justice to animals because we share similar tissues, appetites, emotions, reasonings and perceptions in common. Elsewhere, he also attacked the insensitivity of the Christian Jesus who drove a madman's demons into a herd of helpless

Gadarene swine when he could just as easily have banished them from the Universe.[12]

It appears, then, that early Christianity adopted a distinctly biased, *anti-animal* version of the ancient debate; a version which, in Sorabji's view, 'accounts for the relative complacency of our western Christian tradition about the killing of animals'.[13] This one-sided representation was transmitted through the writings of Augustine, and later consolidated and refined by the medieval Dominican scholar, Thomas Aquinas (1225–74), who not only denied rationality to animals but also immortality. Like Aristotle, on whose works he drew heavily, Aquinas believed that only the reasoning part of the soul survived the body after death. Since animals lacked the power of reason, their souls therefore perished along with their bodies. This apparently simple conclusion had far-reaching implications. By denying animals an afterlife, Aquinas rescued Christians from the otherwise alarming prospect of encountering the vengeful spirits of their hapless animal victims somewhere in the hereafter. It therefore reinforced the notion that humans had no reason to feel morally concerned about the treatment of non-human species. As if to emphasize this point, Aquinas also reinterpreted Old Testament passages that appeared to advocate kindness toward brute beasts:

If in Holy Scripture there are found some injunctions forbidding the infliction of some cruelty towards brute animals . . . this is either for removing a man's mind from exercising cruelty towards other men, lest anyone, from exercising cruelty upon brutes, should go on hence to human beings; or because the injury inflicted on animals turns to a temporal loss for some man.[14]

Elsewhere[15] he repeated himself: 'God's purpose in recommending kind treatment of brute creation is to dispose men to pity and tenderness towards one another.' In other words, according to Aquinas, people had no direct moral duties toward animals whatsoever. Wanton cruelty was to be avoided, but only for economic reasons or because it might encourage cruelty to fellow humans; not because of the suffering inflicted on the animals themselves. Animals had no rights because 'only a person, that is, a being possessed of reason and self-control, can be the subject of rights and duties'.[16] Aquinas also sanctioned Aristotelian physics and astronomy, according to which the Sun and the planets revolved around the Earth which was conveniently fixed at the centre of a finite Universe.[17] Thus, in one fell swoop, he placed Man on a pinnacle at the very centre of creation, and endowed him with a God-given right to absolute mastery over other life-forms.

According to the historian, Keith Thomas, this 'breathtakingly anthropocentric spirit' achieved its highest expression during the sixteenth and seventeenth centuries. Theologians and philosophers of this period tended to perceive God's beneficent purpose in everything. And the undoubted purpose of animals, whether docile or noxious, was to serve mankind, either by providing him with food, raw materials or entertainment, or by labouring for his practical or moral advantage. Their reasoning on such matters was often so contrived as to seem ridiculous by modern standards. It was suggested, for example, that fishes often shoaled near the shore because they were intended for human consumption; that the Creator made horses' excrement smell sweet because he knew that men would often be in its vicinity, and that horse-flies were created so that people could exercise their ingenuity in avoiding them. The supreme arrogance of this viewpoint is exemplified by the following comment of Francis Bacon (1561–1626):

Man, if we look for final causes, may be regarded as the centre of the world, insomuch that if man were taken away from the world, the rest would seem to be all astray, without aim or purpose.[18]

The situation reached its climax in the early seventeenth century in the writings of the brilliant and influential French philosopher, René Descartes (1596–1650). In his *Discours de la Méthode* (1637), Descartes proposed that the bodies of both animals and humans could be adequately compared with complex machines, such as clocks, that functioned according to relatively simple mechanical principles. Humans, however, differed fundamentally from animals in possessing rational

and immortal souls, and the faculty of true speech, whereas
animals were essentially no different from mindless and
soulless automata. From a moral standpoint, the crucial issue
in the Cartesian 'beast-machine' theory was the question of
pain. Although Descartes himself never maintained that
animals were absolutely insensitive to pain, some of his more
zealous followers succeeded in propagating this idea.[19] A
contemporary eye-witness recorded how they:

administered beatings to dogs with perfect indifference, and made
fun of those that pitied the creatures as if they felt pain. They said
the animals were clocks; that the cries they emitted when struck
were only the noise of a little spring that had been touched, but that
the whole body was without feeling. They nailed poor animals up on
boards by their four paws to vivisect them and see the circulation of
the blood which was a great subject of conversation.[20]

The anthropocentrism that characterized medieval and
Renaissance thought and theology was far more than just an
arbitrary ideal. It was a fundamental and fiercely dogmatic
moral precept whose exponents vigorously and sometimes
violently opposed alternative doctrines. Nowhere was this
attitude better exemplified than in the fanatical depredations
of the medieval Inquisition. From the second half of the
twelfth century onwards, the spread of religious dissent in
Europe prompted religious and secular leaders to devise new
ways of identifying and combatting heresy. In the year 1231,*
the Papal Inquisition emerged as the instrument of this
suppression.[21] Although the official purpose of the Inquisition
was to search out and excommunicate obdurate heretics, in
practice, during the Middle Ages and the early modern period,
its main goal appeared to be the eradication of anyone or
anything that contradicted the biased, hierarchical,
Aristotelian/Thomist view of man's place in nature.

The Inquisition's assault was two-pronged. At one end of the

---

* The Papal Inquisition did not become fully organized until the latter half of
the thirteenth century. However, Germany's first official inquisitor, the blindly
fanatical Conrad of Marburg, was appointed with full papal authority in 1231
(Cohn, 1975: 24).

scale, those who questioned the veracity of Aristotle's geocentric or earth-centred Universe were attacked because they inevitably raised doubts about humanity's central position within the Cosmos. Copernicus' treatise *De revolutionibus*, published in 1543, revived the much older theory that the Earth revolved around the Sun rather than vice versa. Fortunately for him, Copernicus died before he had time to face his critics who were numerous and vociferous. Others were less fortunate. Giordano Bruno (?1548–1600) combined the new Copernican doctrine with the idea of an infinite Universe, and concluded that our solar system was neither at the centre of the Cosmos nor unique.[22] He postulated that many such systems might exist, some of them inhabited, and had the temerity to suggest that man was 'no more than an ant in the presence of the infinite'.[23] The Church tried to force him to recant these heresies, but he refused and was promptly burnt at the stake. Ten years later, Galileo Galilei published his *Siderius nuncius* which upheld the Copernican heliocentric theme, and almost met with the same fate. The Roman Inquisition imprisoned him and forced him to recant; a process that left him a broken and disillusioned man.

At the other end of the spectrum, the inquisitors reacted with murderous brutality toward anyone or anything that threatened to undermine the distinction between human and animal, culture and nature. Before the mid-thirteenth century, the Church had adopted a relatively lenient attitude to the variety of pagan religious beliefs that abounded locally throughout Europe.[24] The Inquisition systematically rooted them out and obliterated them. Harmless cults involving nature worship, and superstitious rituals connected with pre-Christian deities or sacred groves, trees, streams and wells were ruthlessly suppressed.[25] Animal cults, such as that of St Guinefort, the greyhound saint (see chapter 6), were extirpated and their adherents dispossessed, excommunicated and, in some cases, executed.[26]

One of the largest and most influential groups of medieval heretics, the Cathars, preached a doctrine which had more in common with Buddhism than Christianity. Like earlier heretical sects, such as the Manicheans and Bogomils, Cathars believed in metempsychosis or reincarnation. The soul, according to Catharist belief, was locked in an eternal cycle of death and re-birth, and could be reborn in either animal or human form. For this reason, all warm-blooded creatures were deemed to possess souls of equal dignity to those of humans. Senior or Elect Cathars abstained from eating meat or any animal products, such as milk, cheese or eggs, and took a solemn vow never to kill any warm-blooded creature, even to save their own lives. Only in this way, they believed, was it possible for the soul to escape the endless cycle of earthly existence and achieve a state of divine grace. As a consequence of their beliefs, the Cathars were the objects of a long and bloody religious and secular crusade. In the Languedoc region of southern France, many thousands were massacred without trial or burned at the stake during the thirteenth and fourteenth centuries. The French, incidentally, nicknamed the Cathars *'bougres'* (literally 'Bulgarians'), from which a common English term of abuse is derived.* In the modern vernacular the word denotes unnatural or bestial sexual relations, although élite Cathars were themselves excessively abstemious and foreswore all forms of sexual activity.[27] The link between heresy and bestiality is interesting, however, because it crops up again in the fifteenth century during the European witch persecutions.

Within the space of roughly two centuries, it has been estimated that as many as three-quarters of a million people were tortured and burned by the Inquisition for alleged witchcraft. In continental Europe, witches and sorcerers were believed to make pacts with Satan during orgiastic rituals, in which the Devil, usually disguised as a large animal such as a goat, a dog or an enormous cat, supposedly engaged in unnatural sexual acts with his disciples.[28] Throughout the

---

* According to Norman Cohn (1975: 57) contemporary commentators even attempted to derive the word 'Cathar' from the Low Latin for 'cat' (*cattus*) since it was in the form of a huge black cat that Lucifer supposedly appeared to them in their ceremonies. In reality, the word is derived from the Greek 'Katharoi' meaning 'the pure ones'.

Renaissance, bestiality – euphemistically referred to in law books as *offensa cujus nominatio crimen est** – was regarded by the Church as the most horrendous and unspeakable of sins. Individuals convicted of the offence were publicly tried and executed along with their animal paramours.[29] Such an extreme reaction to what is, in effect, a relatively harmless activity was not in the least surprising. Bestiality is, after all, the ultimate anti-anthropocentric act. As one seventeenth-century English moralist put it, 'It turns man into a very beast, makes man a member of a brute creature.'[30] For the same reason, bestiality was also a useful propaganda weapon for the Inquisition to level at those who indulged in heretical practices. The mere suspicion that they might engage in such behaviour was sufficient to damn them in the public eye. Witches, incidentally, were also notorious for their ability to transform both themselves and others into beasts.[31]

Pet-keeping also constituted a form of heresy at this time, but for precisely the opposite reason that bestiality was abhorred. Instead of turning humans into beasts, pet-keeping turned beasts into humans or at least semi-humans. As should be apparent from previous chapters of this book, people who keep animals for companionship tend to treat them like members of the family. They give them personal names, talk to them as if they understand much of what is being said, and attribute human thoughts, qualities and emotions to them. They develop attachments for companion animals which are only comparable, in terms of their intensity and duration, to the sort of attachments people form for each other. And when pets die they are often mourned, and occasionally buried with ceremonial honours. A small minority of people even dress their pets up in clothes and pack them off to summer camps. This tendency to anthropomorphize or humanize the animal is an important ingredient of the owner–pet relationship. Without it, it would be difficult to perceive the pet as a true companion, or benefit from its company. Undoubtedly, it was also this characteristic that made pet-keeping, during the

---

* 'the offence the very naming of which is a crime'.

Middle Ages and the Renaissance, a potentially serious threat to the foundations of religious and philosophical belief.

When, in the thirteenth century, William of Wykeham discovered that the nuns of Romsey Abbey were keeping pets, he warned them that such acts imperilled their mortal souls.[32] Similarly, William Harrison accused the nobility of wantonness, corruption, concupiscence and irksome idleness because of their pet-keeping habits, and a later moralist strongly condemned 'over-familiar usage of any brute creature'.[33] Attitudes to the animals which are nowadays normally kept as pets were also very different. As Keith Thomas points out, Middle Eastern views of the dog as a filthy and disreputable scavenger had been transmitted via the Bible to medieval England and were still widespread, if not prevalent, during the sixteenth and seventeenth centuries. In nearly fifty recorded allusions to dogs, Chaucer does not employ so much as a single commendatory or complimentary adjective.[34] Likewise, Shakespeare had nothing nice to say about them, and many preachers of the period used the dog in their sermons as a metaphor for all of man's baser qualities: gluttony, crudity, lust, incestuousness, promiscuity and general disruptiveness.[35] The reputation of the cat was scarcely better. In 1607, Edward Topsell ruled that the cat was 'an unclean and impure beast that liveth only upon vermin and by ravening'.[36]

During the period of the English witch trials, pet-keeping also became an excuse for virulent persecution. English sensitivities evidently baulked at the continental notion of bestial sexual encounters between necromancers and the Devil. Instead, the authorities focused their attentions on the next best thing: sentimental attachments between socially isolated old people and their pets. Barbara Rosen, the author of *Witchcraft*, summarizes English attitudes:

> The element of affection in the alliance, which, on the continent, took the form of surrender and worship, and bestiality with demons was in England expressed by the cosy, slightly perverted relationship of a lonely and poverty stricken woman to her pet animal.[37]

One need hardly say more.

The martyrdom of Giordano Bruno and the persecution of Galileo Galilei in the early part of the seventeenth century marked a turning point in the tide of anthropocentrism that had dominated European thought and philosophy for the previous five centuries. The power of the Inquisition was waning, and Aristotle's earth-centred and, ultimately, human-centred universe soon lay in ruins, thanks largely to the efforts of the astronomers Tycho Brahe, Johannes Kepler, Robert Hooke and Isaac Newton.[38] A few dissenters even began to speak out in favour of animals. Among the first of these was the French sceptic philosopher Michel de Montaigne (1533–92), who argued that cruelty to animals was wrong in itself, irrespective of whether or not it led to cruelty to humans. He also attributed the tendency of human beings to view themselves as separate from and superior to other creatures to 'vanity of the imagination' and the 'disease' of presumption.[39] But to begin with, Montaigne and others like him were relatively isolated voices crying out in a wilderness assuredly dominated by humanity. Another two centuries were to pass before there would be any significant practical changes in the treatment of other life-forms.

The precise course of this change has been thoroughly documented by Keith Thomas and others, and need not be summarized here. Suffice it to say that from about the middle of the eighteenth century onwards, a growing stream of literature – poems, pamphlets, philosophical essays and educational tracts – were published in England, advocating the humane treatment of lower animals.[40] Probably the most influential of these publications was the *Introduction to the Principles of Morals and Legislation*, written by the brilliant moral and political philosopher, Jeremy Bentham, which appeared in 1780. Bentham's argument on the subject of cruelty to animals was simplicity itself. He did not dispute the notion that, in many respects, humans were superior to animals. He merely pointed out that these differences between humans and non-humans were morally irrelevant. The fact that animals had only limited reasoning abilities and lacked the power of speech was immaterial as far as Bentham was concerned. 'The question', he said 'is not, Can they *reason?* nor Can they *talk?* but, *Can they suffer?*'* Ironically, the Cartesian vivisectors had sowed the seeds of their own destruction. All the evidence that they had accumulated on the internal anatomy and physiology of animals merely served to emphasize their similarity to humans. And if the underlying mechanisms and responses were the same, then it was highly probable that animals and humans both experienced similar sensations of pain and discomfort. And if this was the case, then there could be no moral justification for ignoring animal suffering, any more than one should ignore the suffering of an irrational and speechless human infant.[41]

By the turn of the century, concern for animal welfare was widespread among the educated middle classes, and at the beginning of the nineteenth century a series of bills was proposed in the House of Commons to outlaw various forms of cruelty to animals. Initially they were staunchly opposed and defeated, sometimes ignominiously. Peter Singer, in his book *Animal Liberation*, quotes a report from *The Times* of 1821 when Richard Martin MP proposed a law prohibiting the ill-treatment of horses:

when Alderman C. Smith suggested protection should be given to asses, there were such howls of laughter that *The Times* reporter could hear little of what was said. When the Chairman repeated this proposal, the laughter was intensified. Another member said Martin would be legislating for dogs next, which caused a further roar of mirth, and a cry 'And cats!' sent the house into convulsions.

Nevertheless, the following year, Martin succeeded with a similar bill which, as it happens, included asses but neither dogs nor cats.[42] In the same year bull-baiting was made illegal on the public highway, and two years later the Society for the Prevention of Cruelty to Animals (which later received royal

---

* It should be emphasized that Bentham was not the first to base the moral considerability of animals on their apparent 'sentience'. Jean-Jacques Rousseau proposed the same idea in 1755 in his *Discourse on the Origins and Foundations of Inequality among Men*, as did the more obscure English clergyman, Humphrey Primatt, in his *The Duty of Mercy and Sin of Cruelty to Brute Animals* published in 1776 (Tester, 1991: 131; Maehle, 1994: 92–3).

patronage) was founded. Bull-baiting was banned entirely in 1835, as were badger-baiting and dog-fighting. Cock-fighting continued until it was finally prohibited in 1849.[43]

The English were ahead of the field when it came to humane considerations about animals. Writing in the same year as Jeremy Bentham, 1780, the great German philosopher, Immanuel Kant, reiterated the anthropocentric, Thomist doctrine:

> So far as animals are concerned we have no direct duties. Animals are not self-conscious, and are there merely as a means to an end. That end is man. . . If a man shoots his dog because the animal is no longer capable of service, he does not fail in his duty to the dog, for the dog cannot judge, but his act is inhuman and damages in himself that humanity which it is his duty to show towards mankind.[44]

And, not surprisingly, the Catholic Church remained consistently dubious about people's moral responsibilities toward animals. As late as the middle of the nineteenth century, Pope Pius IX refused permission for the establishment of a Society for the Prevention of Cruelty to Animals in Rome because he deemed it a theological error to suppose that man had any duty toward animals at all.[45] In many predominantly Catholic countries, the notion that 'animals have no souls' is still commonly used as a means of justifying indifference to their welfare.*

The growing concern for animal welfare in eighteenth- and nineteenth-century England was accompanied and doubtless promoted by a tremendous increase in the popularity of pets, which seemed to spread downward from the aristocracy and into the then expanding middle classes.[46] A developing interest in wild animals, plants and the mysteries of the

---

* The recent 1994 edition of the *Catechism of the Catholic Church* advocates kindness to animals out of respect for God, their creator, as well as stating that it is 'contrary to human dignity to cause animals to suffer or die needlessly'. However, it also endorses the rigorously anthropocentric view that 'animals, like plants and inanimate beings, are by nature destined for the common good of past, present and future humanity'. Precisely how something can be 'destined' for anyone's 'past' good is not made clear in the text.

natural world also began to emerge at this time. Formerly, nature had been perceived as something vaguely threatening and inimical to human security, and animals, when they were discussed at all, were described almost entirely in terms of their useful or noxious qualities. They had no intrinsic value in themselves, and were certainly not considered worthy of serious study. But with the dawn of the so-called 'Age of Enlightenment', attitudes began to change. Among the first pioneers in the new field of natural history was John Ray (1627–1705), the son of an Essex blacksmith, who devoted his life to the study of nature. He travelled widely, meticulously recording everything he saw, and produced a number of compendious tomes on various groups of animals and plants. Ray's successor was not an Englishman but a Swede; Carl von Linné – alias Linnaeus – went a stage further than Ray and compiled an exhaustive and detailed inventory of all the living organisms he encountered. In it he included the human species, although most of his contemporaries ignored the implications of this and continued to believe in an absolute separation between humans and non-humans. He also devised a method of naming and classifying animals and plants, according to *genus* and *species*, which remains the standard system of biological nomenclature to this day.

Linnaeus was an extraordinarily influential man. All over Europe he instilled others with his own enthusiasm, and at the same time provided them with a universal language for describing and cataloguing living forms. Soon, natural history became a respectable and fashionable pursuit among the gentry and the educated classes. Everywhere people were madly collecting, recording and classifying new discoveries using the handy Linnaean system, and no voyage to the Indies or Africa was complete without its resident naturalist aboard.[47]

John Ray, Linnaeus and their many followers were deeply religious men who never doubted the existence of God. On the contrary, they viewed the workings of nature – all the miraculous and ingenious adaptations of animal structure and behaviour – as irrefutable evidence of the Almighty's benevolent purpose and design. To them, species were unchanging entities. Each was exactly as it had been on the day the Creator brought it into being, and each was superbly designed to fulfil the unique role in life assigned to it. Yet, at about the same time, fossil evidence began to emerge which suggested that species not only changed but also, from time to time, became extinct. And, toward the end of the eighteenth century, this encouraged a few naturalists, such as the Frenchmen, Buffon and Lamarck, and the English polymath, Erasmus Darwin, to revive the old and previously discredited notion of species mutability, the concept that living organisms could change or evolve over time. A minority even attempted to apply the same notion to humanity. The eccentric Scottish sage, Lord Monboddo, for example, asserted that orang-utans were not animals at all, but rather a race of humans that had, in evolutionary terms, missed the boat. He also believed that similar, brutish, pre-linguistic humans inhabited the forests of Angola.[48] Such ideas were perhaps bizarre by modern standards but, nevertheless, it was speculation of this kind which paved the way toward the revolutionary theory of evolution by natural selection, formulated independently by Charles Darwin and Alfred Russel Wallace.

In many respects, Charles Darwin's *The Origin of Species*, published in 1859, was the last great work of heresy. Had it appeared two or three centuries earlier, both the book and its author would doubtless have gone up in smoke. As it was, the response to publication, particularly by the Church and the less well-educated social classes, was overwhelmingly negative.[49] The crux of the Darwin/Wallace heresy was the idea that all organisms and, by inference, human beings (although Darwin wisely refrained from a discussion of human evolution until 1871) were the products of a natural and essentially random process. At any given time, and in any given environment, some individual members of a species possessed physical or behavioural characteristics that enabled them to survive and reproduce more successfully. In the 'Struggle for Existence', as Darwin called it, these individuals inevitably prospered and proliferated at the expense of others less well endowed. They also passed their winning qualities on to their offspring who in turn prospered, as long as conditions favoured them. Variation was promoted by random changes, but apart from this, no supernatural guiding or propelling force needed to be invoked in order to keep the ball, so to speak, rolling.

The most important aspect of this theory was that it was entirely devoid of the concept of *design*. It made no use of teleology or arguments couched in terms of *purpose* or *final causes* or *progress* toward some higher goal. In this respect it was fundamentally different from all of its essentially Aristotelian predecessors, and it was this aspect, above all, that enraged the establishment. Eighteenth- and nineteenth-century theologians had, until then, used *design* as one of their principal arguments in favour of the existence of God. Whether or not species evolved, animals and plants were seen as having been custom-built by the Creator to carry out their specific purposes in life. More often than not, these purposes were to serve humanity.[50] With the concepts of purpose and design gone, the whole idea of a supreme creative being or God seemed almost superfluous, as did the notion of human superiority over other species. All at once, humans, animals, plants and other organisms were flung together in the same boat, driven onwards by the inexorable pressure of natural selection.

Although both Darwin and his theory were ridiculed in the press, and denounced by many eminent Victorians – who took particular exception to the idea of being descended from apes – *The Origin* was a best-seller and was enormously influential. Its strength lay both in its logical consistency and in the sheer weight of evidence which Darwin had amassed in support of his ideas. Darwin was also fortunate in having a number of powerful advocates, such as Thomas Huxley, who never missed an opportunity to expound the new theory publicly, while the author himself scrupulously avoided the limelight.[51] Eventually, even the Anglican Church bowed to Darwin's doctrine; first, by admitting that the biblical story of creation was purely allegorical and, second, by reconciling itself to the possibility that God had merely set the wheels of evolution in motion; natural selection had done the rest.[52] Within the past century, the triumphal progress of Darwinian theory has continued, and it is now accepted orthodoxy in most western countries and throughout the scientific world. Predictably, this trend has been accompanied by a sharp decline in traditional religious beliefs, and by a continually growing enthusiasm for the biological sciences, conservation and environmentalism, animal welfare, pet-keeping, and other non-anthropocentric pursuits.

It would, however, be a great mistake to imagine that anthropocentrism is now dead and buried. In the United States, resistance to the Darwinian revolution is still widespread. In 1925, the State of Tennessee passed a law which banned the teaching of biological evolutionism in public schools. This law was only repealed in 1967. Since then, there has been a resurgence of Christian fundamentalism in North America, with far-right organizations such as the 'Moral Majority' successfully campaigning to have so-called 'scientific creationism' taught alongside the theory of evolution in schools.* The Catholic Church, from the start, has adopted a largely fundamentalist position on the subject of evolution. Its first and only concession to the theory appeared in 1951 when Pope Pius XII ruled that 'People should not take it for granted that evolution is a proved fact and should not act as if there were no theological reasons for reserve and caution in their discussions.'[53]

The odour of anthropocentrism also lingers in unexpected areas of the natural sciences. In 1894, Lloyd-Morgan, one of the founding fathers of the study of animal behaviour, introduced his famous canon which stated that:

In no case may we interpret an action as the outcome of a higher psychical faculty, if it can be interpreted as the outcome of the exercise of one that stands lower in the psychological scale.

---

* In 1995 the State of Alabama School Board passed a ruling that all student biology textbooks should contain a clear statement that evolution is only one of several theories concerning the origin of species.

This was generally interpreted to mean that we should never attribute higher thoughts or feelings to animals, when their behaviour can be explained in terms of simple reflexes and instincts. This essentially arbitrary dictum introduced a systematic anthropocentric bias to the study of animal behaviour which has remained there ever since. Nowadays it is justified in the interests of scientific objectivity, and a desire to avoid assuming that animals necessarily experience the same thoughts and feelings as ourselves. On the face of it, this may seem a sensible and harmless rule of thumb, but it has undoubtedly had the effect of reinforcing the barriers between our species and every other. Anthropomorphism has become a dirty word, and the application of terms such as happy or dejected, or surprised or bored to animals is regarded as totally unacceptable, whether or not their behaviour is consistent with these emotions.[54] Yet, as the scientist Donald Griffin points out, 'it is no more anthropomorphic, strictly speaking, to postulate mental experiences in another species than to compare its bone structure, nervous system, or anti-bodies with our own'.[55] But, instead of giving animals the benefit of the doubt, scientists consistently employ euphemistic and often unnecessary jargon when describing their behaviour, and may even resort to putting words such as pain or hunger in quotes, as if it were doubtful whether non-humans could experience even these basic sensations.[56] Meanwhile, at the other end of the scale, issues such as language acquisition in the great apes, consciousness in animals, or the application of Darwinian theory to the evolution of human culture and behaviour continue to arouse vehement controversy. Notwithstanding the demands of scientific objectivity, it is difficult to escape the conclusion that human beings are still extremely reluctant to admit that the line which separates them from other species is both tenuous and fragile.

In short, old habits die hard. For well over 2,000 years, European religion and philosophy has been dominated by the belief that some omnipotent supernatural being placed humanity on a moral pedestal, high above the rest of creation.

From this vantage point we exercised absolute mastery over other living beings, and even believed that their sole *raison d'être* was to serve our own selfish interests. The great revolutions in science within the past 500 years have been those that challenged and undermined this view and, one by one, they have gradually succeeded in toppling us from our once lofty perch. But we did not go down without a fight. Throughout history the Church or the secular establishment has fought viciously to maintain the *status quo* and to uphold the myth of human supremacy. Nowadays the vestiges of this struggle remain in half-hearted attempts to deny or repudiate our kinship with other species, and our moral responsibilities for their welfare. Part and parcel of this denial, I would argue, has been the lingering tendency to dismiss close, affectionate relationships between humans and animals as trivial, wasteful, perverted and wholly unworthy of serious consideration.

The spread of pet-keeping in the western world may or may not have something to do with rising living standards, or demographic changes in family and community relationships. But it is clear that since the Middle Ages the growth in popularity of companion animals has been inextricably linked with the decline of anthropocentrism, and the gradual development of a more egalitarian approach to animals and the natural world.

CHAPTER 10

Killer with a conscience

the primordial guilt of life that lives on life.
Joseph Campbell, *The Way of the Animal Powers*

The whole idea of human moral supremacy was a myth contrived from an odd mixture of biblical and classical sources which achieved formal expression in the writings of Thomas Aquinas during the thirteenth century. Despite various modifications and setbacks, it dominated western belief for the following 700 years, and its power has only recently declined under the sheer weight of scientific and philosophical evidence to the contrary. Religious and secular authorities succeeded in perpetuating this myth by staunchly resisting attempts to put forward an alternative view of man's place in the universe. Yet it is not immediately obvious why they went to so much trouble and effort to sustain a system of belief that ran contrary to observable facts. Doubtless the concept of human dominion appealed to European egotism but, on its own, this would hardly have provided sufficient justification for the horrors of the Inquisition, or the wholesale persecution and torture of hundreds of thousands of innocent and frequently right-minded people. So why, then, was this particular myth defended and upheld with such ferocious determination? Intriguingly, the inventor of the infamous 'beast-machine' theory supplied a plausible answer to this question more than three centuries ago.

René Descartes was quite explicit in recognizing the practical advantages of his mechanistic doctrine. He saw it not as cruel to animals but rather as 'indulgent to men . . . since it absolves them from the suspicion of crime when they eat or kill animals'.[1] In other words, if one believed that animals, like people, had souls and could suffer, then one would be plagued with guilt every time it was necessary to slaughter them, even if such suffering was ordained by God. But if one accepted the Cartesian view that animals were soulless, insensate machines, then one could do what one liked to them without any moral compunctions whatsoever. The early Christian (and Aristotelian) view that animals were created purely for the benefit of mankind, and the Cartesian idea that they were incapable of suffering were mutually compatible variations on the same theme. Both provided human beings with a licence to kill; a permit to use or abuse other life-forms with total impunity. I am indebted to Keith Thomas for summarizing this argument:

In drawing a firm line between man and beast, the main purpose of early modern theorists was to justify hunting, domestication, meat-eating, vivisection (which became common scientific practice in the late seventeenth century) and the wholesale extermination of vermin and predators.[2]

According to Thomas, the myth that humans were entitled to lordship over the rest of creation was a useful cultural adaptation that greatly facilitated agricultural and economic expansion. It allowed domestic animals to be regarded as objects and merchandise, and it encouraged an aggressive, exploitative attitude to the natural world. Wild animals which were deemed to be useless, or which made the mistake of competing with man on his own ground were universally classified as vermin that needed to be exterminated at every possible opportunity. And uncultivated areas, such as forests, moorlands and heaths, were viewed as bleak and hostile wildernesses that harboured blood-thirsty wolves and legendary monsters. It was man's duty to tame such areas; to subjugate them and bring them under the yoke of human domination.[3] In other words, the myth was important, and was defended so vigorously, because it had immense survival value. At the time, the vast majority of people lived in the countryside and worked on the land. Existence was harsh and uncompromising. The food supply was precarious, life expectancy was short, and farming involved an unending battle with adverse weather, and the constant encroachment of weeds, pests, predators and disease.[4] Moral scruples about the welfare of animals or the destruction of nature would have been singularly out of place in such a world. The rural peasant population had enough problems without burdening itself still further with a guilty conscience. Far easier to accept a blanket doctrine that absolved them entirely from any sense of responsibility for their actions toward other species.

This idea makes a great deal of sense, and it also draws attention to another characteristic of human beings which distinguishes them from animals. A dog, for example, has no need for self-justification when it kills a rabbit, and a cat shows no signs of remorse when it plays with a half-dead mouse. Such suffering is in the nature of things; an inevitable outcome of the relationship between a predator and its prey. For humans, however, matters are not that simple. Although there are exceptions, people generally find it difficult to kill or harm other animals with total indifference. Somewhat ironically, this inhibition seems to arise from our inability to differentiate clearly between animals and ourselves.

Many animals, especially in captivity, will begin to treat members of other species as if they belonged to the same species as themselves. They may, as a result, regard them as potential social or sexual partners, territorial rivals and so on. In a much publicized case, for instance, a captive female gorilla adopted a kitten as a companion, and showed obvious signs of distress when her pet was subsequently run over by a car.[5] Comparable stories involving an enormous variety of other species have also been widely reported.[6] Humans exhibit a similar capacity to generalize beyond the boundaries of their own species, only for some reason this tendency is exceptionally exaggerated and all-inclusive.

The habit of attributing human characteristics to animals and other living and, occasionally, lifeless things is commonly referred to as anthropomorphism. Although it can be controlled or suppressed when necessary, anthropomorphism appears to be a universal human proclivity; so much so that scientific and educational literature is riddled with solemn warnings about its hazards.[7] Up to a point these caveats are justified since the ill-informed use of anthropomorphism can lead to serious misunderstandings of animal behaviour. In 1961, for example, when a chimpanzee called Ham returned from a suborbital space-flight, he was photographed peering out of the capsule and grinning broadly at his rescuers. This image was subsequently reproduced on the cover of *Life Magazine* with a caption implying that he must have enjoyed the flight. Only later did experts on chimpanzee behaviour point out that these animals normally 'grin' when they are frightened. Ham had almost certainly found the whole experience terrifying.[8] Arguments against the use of anthropomorphic interpretations are mentioned here not because they have any obvious bearing on what follows, but because they underline the fact that anthropomorphism is the normal and immediate response of the vast majority of people to animals. If it was not, such warnings would be unnecessary.

Adults can be trained to think about animals and animal behaviour in objective, non-anthropomorphic terms, but children seem to find this much more difficult. During the first years of life, they do not appear to be able to make a clear distinction between humans and non-humans, and even as early as two years of age will begin responding socially toward animals such as family pets and treating them, to all intents and purposes, as if they were persons.[9] Many observers have also noticed that older children can readily relate to real or imagined feelings in animals, when they often have great difficulty relating to or comprehending the feelings of other people.[10] This fact is clearly recognized by the authors and publishers of children's literature who frequently use anthropomorphic animal characters, rather than more realistic images, as a medium for conveying social values and rules.[11] Such observations are subject to a variety of interpretations, but one possible reason why children find it so easy to relate to animals is simply because they are not yet fully indoctrinated with all the paraphernalia of culture, and are therefore more animal-like themselves. They can identify freely with the feelings and needs of animals, whereas the motives and desires of human adults may be quite beyond the realms of their own personal experience. This may also be one reason why so-called 'primitive' peoples often display such exaggerated zoocentric tendencies. Living, as they do, closer to nature, the idea of a firm dividing line between animal and human must appear to them manifestly spurious.[12]

Why human beings should exhibit this all-embracing tendency to humanize or personify animals is something of a mystery, although most writers on the subject assume that it is a natural extension of the same process that enables us to understand and empathize with each other.[13] The psychologist, Nicholas Humphrey, has suggested that what he calls 'reflexive consciousness' – that is, the ability to introspect and reflect upon our own motives and reasons for doing things – is, first and foremost, a social adaptation. It has evolved because it allows us to use self-knowledge in order to predict how others would behave in similar circumstances.[14] If this is the case, then it logically follows that we should use precisely the same criteria to judge and predict the behaviour of non-humans, since they are obviously similar to us in a great many respects. Obviously the technique will give rise to misinterpretations on occasions, just as it can lead to misunderstandings of other people. But used sensibly as a general guideline, it is an extremely effective tool for investigating animal psychology, and one that is widely and successfully employed by the majority of people who live and work with animals. To quote some well-known field primatologists: 'Anthropomorphizing *works* . . . attributing motives and strategies [to monkeys] is often the best way for an observer to predict what an individual is likely to do next.'[15] And even the most experimentally biased psychologists and animal behaviourists, with their avowed antipathy for anthropomorphism, occasionally admit to its effectiveness. John

Garcia, well known for his experiments on learning in rats, openly confessed to using: 'anthropomorphism and teleology to predict animal behaviour because this works better than most learning theories. I could rationalize this heresy by pointing to our common neurosensory systems or to convergent evolutionary forces. But, in truth, I merely put myself in the animal's place.'[16]

It appears, then, that humans have evolved a method of understanding and getting to know other individuals based on personal insight. We are, according to Nicholas Humphrey, 'natural psychologists' with the power to penetrate and explore other minds by the light of our own subjective experience.[17] Ordinarily we use this technique for interpreting and anticipating the behaviour of fellow humans. But because we are animals, and share many of our feelings and motives with animals, we can also use it for comprehending and predicting the activities of non-humans. The trouble is that in doing so we unwittingly transcend the barriers that separate our species from others. And once the barriers are down; once the animal has been, as it were, *personified*, we are under considerable pressure to treat it as if it were, indeed, a person. If we were capable of being truly objective about animals, we would not have the slightest qualms about damaging or destroying them, except, conceivably, for abstruse intellectual or aesthetic reasons. But because we anthropomorphize them and think of them in human terms, we are bound by the same code of morality that governs our interactions with other people. Except in exceptional cases, this has not prevented us from exploiting animals, but it has given rise to a profound moral conflict in our relations with other species. In our efforts to come to terms with this conflict, we have become horribly entangled in an extraordinary web of myths, rituals, fabrications and falsehoods, of which the Christian idea of human moral supremacy is just one of many.

Of course, the nature and extent of this conflict varies according to a variety of circumstances. Much depends, for example, on the particular species of animal involved. For obvious reasons, animals that resemble human beings, either physically or behaviourally, such as penguins, pandas, seal pups, monkeys, dogs, cats and many other 'higher' vertebrates, evoke inordinate amounts of sympathy.[18] They are easy to anthropomorphize, and therefore relatively difficult to exploit with impunity. Conversely, comparatively little concern is displayed for the plight of herrings or laboratory rats, presumably because they possess fewer qualities with which humans can identify. The manner in which animals are exploited is also crucial, since not all methods of employing them are necessarily contrary to their own best interests. The human–pet relationship provides a good illustration. The pet lives in the owner's home, participates in family life as an equal or near equal, and is given a personal name to which it learns to respond. It is cherished during its lifetime and mourned when it dies. The close and intimate relationship that develops between human and animal in this situation is, more often than not, mutually rewarding to both parties, so little if any conflict arises. The same is more or less true in situations where dogs are used for hunting or for herding sheep. The dog, after all, is descended from a wild predator, and it therefore shows a natural inclination to chase or hunt other animals. Dogs do not need to be forced to do these things, although they do require discipline and training to perform the tasks well. In other words, the aims and objectives of the hunter or the shepherd and his dog are roughly compatible. The animal seems to enjoy the work, so the person has little reason to feel guilty about using it.[19] It is primarily in situations where human beings inflict suffering or death on animals that the most serious problems tend to arise.

Hunting is probably the most ancient form of animal exploitation. Human beings have been doing it for at least a million years. To be really successful at hunting, one needs to know a great deal about the habits of the quarry; the kinds of places where it is likely to make its den, the territory it occupies, its seasonal movements, what it likes to eat, where it drinks, when it breeds, how it behaves when threatened, and how best to approach it without causing alarm. One needs in a sense to get inside the animal and see the world from its point of view; to empathize with it. The problem is that empathy often leads to sympathy, and sympathy is in direct conflict with the object of the exercise which is to kill and eat the animal. It is no accident that many former hunters and field-sportsmen have hung up their guns and become ardent conservationists or exponents of animal welfare.[20] The act of getting to know the animal, of becoming personally involved with its feelings, motivations and needs, creates moral obligations that inhibit the ultimately violent purpose of the hunt.* Such inhibitions are widespread in the majority of traditional hunting cultures. Indeed, one authority states that every hunting society is 'dominated by the hopes and the fear connected with the killing of game'.[21]

Typical hunters and gatherers do not view the animals they hunt as in any way inferior to themselves. They are seen as mental and spiritual equals or even superiors, capable of conscious thoughts and feelings analogous, in every respect, to those of humans.[22] This, incidentally, is one possible reason why pet-keeping is so popular among such societies (see chapter 4). Since they do not believe that animals are inferior, no stigma is attached to adopting them as social companions.[23] The sense of equality between humans and non-humans in hunting cultures is often affirmed by the almost Darwinian idea that people and animals are related by descent from a common ancestor.[24] In some cases, kinship with other life forms is formalized in a system of religious belief, commonly known as *totemism*, in which the family, clan or tribe traces its origin back to some mythical animal progenitor.[25] Elsewhere, the individual hunter's affinity with animals may be expressed through the acquisition of animal 'guardian spirits'. Such

---

* One of the best known victims of this process was Aldo Leopold, one of the great intellectual beacons of the American environmental movement. According to his own account, Leopold appears to have undergone a sort of partial conversion after watching 'a fierce green fire dying' in the eyes of a wolf that he had shot. The experience did not stop him hunting, but it completely altered his perception of predators and their vital role in the ecological balance (*A Sand County Almanac*, 1968: 130).

spirits are said to act as mentors and as sources of supernatural power, able to assist the hunter in his daily activities
in return for compliance with strict rules of respectful conduct
toward animals and the natural world.[26] These highly anthropomorphic perceptions of animals provide hunting peoples
with a conceptual framework for understanding, identifying
with, and anticipating the behaviour of their prey. They are
crucial to their success as hunters. But they also generate a
moral conflict because, if animals are believed to be essentially the same as persons or kinsmen, then killing them
constitutes murder, and eating them is equivalent to
cannibalism.

In general, the extent of the guilty feeling that results from
hunting seems to vary from culture to culture. The anthropologist, Joseph Campbell, has argued that it tends to be less
pronounced in societies that live in very stable environments,
such as tropical and equatorial rainforests, while it is often
most exaggerated in those that inhabit less predictable
regions where the conditions of life are more severe.[27] This
variation seems to make sense. In extreme climates, such as
those of desert or arctic regions, the population of game tends
to fluctuate widely from year to year, and the hunters who
inhabit these areas generally display considerable anxiety
about nature, and the forces that influence these changes.
When things go wrong; when the game suddenly vanishes
without warning, they naturally enough look for an obvious
cause to which they can attribute the catastrophe, and
which can then be avoided in the future. More often than not,
in these situations, they assume that the fault lies with
themselves; that the loss or disappearance of the animals
represents a punishment for some moral transgression on
their part.[28] In this context, the pre-existing sense of guilt they
feel about killing animals often appears to become the focus of
concern. It is assumed that some supernatural agency is either
seeking retribution for the animal's death or chastising the
hunter for his failure to show the animal appropriate ritual
respect. The dilemma is eloquently expressed in the following
statement by an Iglulik Eskimo:

The greatest peril in life lies in the fact that human food consists entirely of souls. All the creatures that we have to kill and eat, all those that we have to strike down and destroy to make clothes for ourselves, have souls, like we have, souls that do not perish with the body, and which must therefore be propitiated lest they should avenge themselves on us for taking away their bodies.[29]

In contrast, the hunting and gathering inhabitants of tropical rainforests, such as the pygmies of Zaire, seem to display less of this anxiety. Their world is one of continuous 'vegetal abundance' in which there is scarcely any variation from year to year in the availability of game and other forest foods. Even if they feel guilt about killing animals, they show very little, and never show the same need to compensate the animal for the harm that has been done. Few environmental fluctuations and catastrophes plague their existence and, consequently, they have less need to make amends through ritual acts of restitution.

Another factor that may contribute to the sense of guilt is the degree to which the hunter is economically dependent on, and therefore identifies with, a particular species of prey. In the rainforest, where species diversity is high, the pygmies do not rely heavily on any particular animal, but simply take whatever they can catch. Their hunting methods are also relatively crude, and often involve simply driving every animal in the vicinity into lines of waiting nets.[30] In more extreme environments, however, where the diversity of suitable game animals is relatively low, hunters may find themselves utterly dependent on only one or two species. And in such areas, the animals tend to acquire an overwhelming economic, social and mystical significance. The Naskapi caribou hunters of the Canadian arctic provide a good example. From earliest childhood, every Naskapi is immersed in the lore and legend of caribou. He is regaled with stories of caribou psychology and caribou behaviour; he is given advice on tracking them, stalking them and killing them, and he listens to stories of previous hunts.[31] By the time he reaches adulthood, he will know all there is to know about caribou, and he will identify with them as closely as if they were family or friends.

Inevitably, with this level of emotional involvement between hunter and prey, the act of killing has much of the flavour of homicide, and necessitates the evolution of elaborate acts of restitution and atonement.

According to Naskapi mythology, the caribou possess a spiritual guardian, variously known as the 'Animal Master' or 'Caribou Man'. Sometimes he is conceived of as an anthropomorphic being, but more often he is represented as a single animal in a vast supernatural herd; immortal and indestructible; an archetype of the species. The Animal Master provides the hunter with caribou, but he is easily offended, and has the power to take them away if the hunter proves himself unworthy of the gift. This arrangement was established long ago between the Animal Master and a great shaman. At the time, people had nothing to eat except roots and berries, and were starving, so the legendary shaman set out to look for alternative foods. On his journey he encountered the Animal Master, and so impressed him with his magical powers that he agreed to release the caribou to be hunted. But the covenant was made on the understanding that the hunters would invariably treat the animals with respect and never insult their spirits through acts of arrogance or ridicule. They were also required to eat every morsel of flesh and waste nothing, and to perform regular ceremonies to commemorate the agreement. As long as all these things were done properly, the caribou spirits would return safely to the Animal Master and be continually renewed. But any violation of the rules would cause their disappearance, and starvation and death would then ensue.³²

Remarkably similar beliefs are found among the reindeer hunters of Siberia. Again there is a spirit 'Being' who regulates and controls the supply of game, and reindeer themselves are credited with powers of reasoning and speech. Although the animals are perceived as willing victims conniving in their own slaughter, great care must be taken in the preparation of kills to avoid offending the reindeer's spirit or jeopardizing the future supply of game. The reindeer's death is a rite of renewal as well as an act of destruction. The hunter receives the animal's physical substance – its meat, hide and bone – but its spirit is immortal and undergoes an eternal cycle of death and rebirth.[33]

Although it varies in detail from place to place, the undercurrent of guilt and the need for some form of atonement for animal slaughter is widespread, if not universal, among hunting peoples. In certain African tribes, for example, hunters are obliged to undergo ceremonial acts of purification in order to remove the stain of murder from their consciences. In others, the hunter will beg the animal's forgiveness so that it does not bear a grudge. Rural villagers in Malawi believe that hunters can even be afflicted with a kind of *disease* brought on by the vengeance of certain prey species, especially kudu and eland. Consequently, Malawian hunters take great care to propitiate the spirits before hunting, as well as using protective medicines of various kinds.[34] The Barasana Indians of Colombia regard the act of killing animals as spiritually dangerous, and believe that their flesh is poisonous unless ritually purified first. Generally speaking, their anxieties about killing and eating game increase the larger and more anthropomorphic the animal involved.[35] Among the Moi of Indochina, expiatory offerings are made for any animal killed by hunters, because they believe that it has been taken by force from its spiritual guardian who may decide to seek revenge. Moi folklore abounds with cautionary tales of evil befalling hunters who have failed to make the necessary restitution. Significantly, if an animal is caught in a trap, no offering is required because the Moi believe that the animal's spirit guardian must have deliberately pushed it into the trap, in order to punish it for some misdeed.[36] Similarly, the Chenchu hunters of India propitiate the spirit world for any animal killed by themselves, but consider it unnecessary if the prey is killed by their hunting dogs; presumably because the blame rests with the dogs not with the hunters.[37] The remains of animals after they have been eaten are also treated with ritual respect in many cultures. The bones are often carefully collected, reassembled in something like their original order, and provided with a decent burial.[38] In short, the animal that is killed for food is almost never regarded as a mere object; a passive victim of human predation. On the contrary, it is seen as an active and, hopefully, willing agent of its own slaughter. The successful hunter achieves his goal, not primarily through practical knowledge or skill, but rather by virtue of his respectful attitudes and behaviour toward the quarry. Only then will the animals consider him worthy of the gift of meat and allow themselves to be killed.[39]

Even modern sport hunters in western countries sometimes appear to adopt comparable myths and rituals. Ideally, the hunter undergoes a sort of self-imposed, ritual penance by making things as difficult and uncomfortable for himself as possible. Shooting the easy prey or the proverbial 'sitting duck' is considered bad form, and the greatest satisfaction – and the least guilt, one suspects – is invariably derived from the clean kill in difficult circumstances, preferably after hours or days of personal deprivation and hardship.[40] The modern huntin'-'n'-shootin' mentality, with its class-conscious emphasis on fair play, is probably a nineteenth-century affectation of the English public school system.[41] But the basic notion of sporting conduct in hunting is of much greater antiquity. Plato in his *Laws* for instance, strongly condemns cruel, lazy or deceitful methods of hunting, such as the use of nets, traps or night-stalking, and recommends, instead, the arduous pursuit of land mammals on foot with only the assistance of dogs.[42]

The hunter's need to compensate either the animal itself or the forces of nature is also ancient. The Palaeolithic cave paintings that have been discovered all over Europe at sites such as Lascaux in France or Altamira in Spain may represent the traces of restitutional cults. According to one authority, 'whenever an animal was killed, his essence was restored to Nature by ritual rendering of his image at a sacred spot'.[43] Joseph Campbell has suggested that such practices may be as old as the recognition of death itself. The oldest known ritual burial of human beings was practised by Neanderthal Man roughly 60,000 years ago. Several examples have been found; the most remarkable from a cave in the Zagros mountains of

Iraq, where the skeleton of a man was found whose body had been carefully laid to rest on a bed of evergreen boughs heaped with flowers. Campbell has argued that it was during this period of human evolution that people first became aware of the significance of death, and attempted to rationalize it as a 'passing on' to the next world. It has been suggested that the particular individual buried in Iraq may have been an important medicine man or shaman, since the flowers that covered his death bed were predominantly medicinal herbs that he would have needed to continue his work in the afterlife.[44]

From about the same period, a series of high mountain caves have been discovered in Switzerland and Germany in which the bones and skulls of extinct cave bears appear to have been carefully arranged and stored by Neanderthal hunters. Campbell supports the view that these remains indicate the existence of a Palaeolithic bear cult, and that the bones were kept for the purpose of venerating the spirits of animals that had been killed for food. The fact that the cult coincided with the earliest known human burials suggests that the people of this period did not draw any clear metaphysical distinction between the death of a person and the death of an animal. In Campbell's own words:

the interpretation of death as but a passing had been applied, not only to the subject, man, but also to the objects of his hunting, and, lest resentment on their part should follow upon a good hunting day and so perhaps spoil the next, rites of gratitude, praise and appeasement were enacted.[45]

Until recently, bear cults still survived throughout the northern hemisphere, and all of them were associated with the most elaborate myths and rituals.[46] There are at least two probable reasons for this degree of elaboration. First, bears are highly prized as prey animals. They represent an extremely valuable source of meat, fat and fur, because they are among the largest and heaviest terrestrial mammals available for hunting. Second, bears are exceptionally anthropomorphic. On occasions, they walk upright like gigantic furry

people, and a skinned bear carcass apparently looks discon-
certingly like a human corpse in physical proportions. Even
the modern cult of the teddy bear testifies to their endearing
human-like qualities. In other words, bears are both
economically important and easy to personify, so the conflict
between exploitation and sympathy is particularly intense.

In the bear cults, the animal is regarded as a spiritual
kinsman, as a divinity in his own right, and as a sort of animal
chieftain; a King of Beasts. He is treated with grave respect,
and everything possible is done to avoid offending him,
including never actually referring to him by name. Instead, he
is discussed using a variety of nicknames, such as Cousin,
Elder Brother, Grandfather, Angry One, Old Man, Old Man in
a Fur Coat, That Big Hairy One, Venerable One, Uncle of the
Woods or That Divine One Reigning in the Mountains. The
killing of the bear is a sacred act that must be performed in a
special way. Modern weapons such as guns are eschewed in
favour of traditional knives, spears and clubs so as to make
things more difficult and hazardous for the hunter. The bear
is also tackled in a fair stand-up fight, rather than taking it by
surprise, in order to avoid inciting its posthumous revenge.
Once the bear is dead, it is customary to apologize effusively
to the corpse, and explain the reason for its killing. In some
cases, the hunter will disclaim responsibility by blaming
someone else. Siberian bear hunters, for example, will say:
'Grandfather, it wasn't I, it was the Russians, who made use of
me, who killed you. I am sorry! Very sorry! Don't be angry with
me!'

The consumption of the bear's flesh is also a sacred ritual
that is conducted according to strict rules. Young women and
girls are obliged to conceal themselves to avoid causing
offence, and dogs are ejected from the proceedings lest they
defile the corpse by licking its blood. It is generally considered
necessary to consume the entire animal at one prodigious
sitting so that none of the flesh is wasted, and the entire
performance is conducted in silence, apart from the occasional
chanting and drumming of shamans. Often the bear's head
with the skin attached is given an honorary position at the

feast, and it is offered a dish of its own flesh to consume as an additional token of respect.

The most elaborate and contrived rituals of all are found among the Ainu people of Japan. According to the Ainu, the bear is not really an earthly being at all; he is a temporary visitor from the spirit world whose ultimate objective in life is to return there. The Ainu assist the bear to perform this metamorphosis by killing it. Indeed, their word for 'sacrifice' means literally 'to send away'. This curious mythical embellishment seems to be necessary because the Ainu not only kill wild bears, they also capture and rear bear cubs and then slaughter them.

Throughout its stay within the community, the cub is treated as an honoured guest. It is adopted by a family, suckled at the mother's breast, and treated like any other small child. When it is too large to be handled safely, it is kept in a strong wooden cage and fed on a diet of fish and millet porridge until it is about two years old, and ready to be released from its body and returned to the spirits. At the appropriate time, the bear is gently informed of the good news in the following terms:

O Divine One, you were sent into this world for us to hunt. Precious little divinity we adore you; hear our prayer. We have nourished you and brought you up with care and trouble, because we love you so. And now that you have grown up, we are about to send you back to your father and mother. When you come to them, please speak well of us and tell them how kind we have been. Please come to us again and we shall do you the honour of a sacrifice.

There then follows an elaborate sacrificial ritual in which the animal is led through the village, taunted, strangled and eventually shot through the heart by a marksman's arrow. The people then feast on the bear's flesh for several days, if necessary, until every morsel of the deity has been consumed. Meanwhile, the woman who suckled the cub, and others who have suckled them in the past, display their ambivalence about what has happened by alternately laughing and crying.[47] (Taunting the victim before sacrifice is a widespread ritual observance in many societies. It probably helps to distance the killers both emotionally and symbolically from the animal.)

The important thing to note about the Ainu bear cult is that the bear is no longer a wild animal. It has been tamed and domesticated, adopted by the community, and thoroughly socialized to people. Doubtless, the practical purpose of the exercise has been to fatten the bear for slaughter, but the bond of sympathy that has inevitably developed between the animal and its keepers renders its sacrifice and consumption metaphorically equivalent to cannibalism. The Ainu appear to have come to terms with the moral contradictions inherent in this relationship by means of self-deception. They have fabricated a system of religious belief that absolves them of guilt for the killing. They have even persuaded themselves that they are doing the animal a service by helping to reunite it with its spiritual parents. This perception of the animal's death is subtly and yet profoundly different from that which is typical of the majority of hunting societies. In most, the guilt is acknowledged and accepted as an unpleasant but unavoidable burden that can only be expiated through acts of penance, reverence and ritual respect. The Ainu, however, like the early Christians, have found a way of making a virtue out of necessity. They have devised an ingenious fantasy that enables them to nurture and care for an animal, and yet slaughter and eat it with a clear conscience. Self-justifying fantasies of this kind are the rule rather than the exception in societies that depend on the exploitation of domestic species.

CHAPTER 11

Licensed to kill

So convenient a thing it is to be a reasonable Creature, since it enables one to find or make a Reason for everything one has a mind to do.

Benjamin Franklin, *Autobiography*

The moral dilemma confronted by farmers and livestock herders is far more serious than that faced by hunters of wild game. Traditionally, successful animal domestication and husbandry has always depended on the shepherd, farmer or stockman having a detailed understanding of his charges. Ideally, he or she will need to get to know them, not just as a group, but as individual personalities, each with distinctive moods, characteristics and temperament. He will need to use his imagination to anticipate the sorts of things that will frighten or attract them, and he will need to be so familiar with them that he can recognize immediately when they are feeling out of sorts. A traditional farmer who neglected to do these things would endanger his own life and livelihood as well as those of his animals. The philosopher, Mary Midgley, sums this up precisely:

People who succeed well with them do not do so just by some abstract, magical human superiority, but by interacting socially with them – by attending to them and coming to understand how various things appear from each animal's point of view. To ignore or disbelieve in the existence of that point of view would be fatal to the attempt.[1]

The ethical problem is more serious for the farmer than the hunter because the relationship with the animal is different.

The hunter may possess remarkable insight into the animal's character, but he never has the opportunity to interact with it socially, and therefore has little chance of developing specific attachments for particular individuals. Moreover, except at the moment of the animal's death, the hunter exercises little control over it. The animal remains an independent being with a mind of its own. In contrast, the domestic animal is partially or wholly dependent for survival on its human custodian, and it will, if given the chance, learn to recognize him and treat him as a social partner. Similarly, unless he takes steps to prevent it happening, the farmer or stockman will get to know individual animals and may become personally attached to them. Once this has happened, the slaughter or the deliberate infliction of suffering on the animal inevitably generates feelings of guilt and remorse because, in human terms, it constitutes a gross betrayal of trust. Farmers, husbandmen and others who benefit from the harmful exploitation of domestic species, have learned to cope with this dilemma using a variety of essentially dishonest techniques. Unfortunately, owing to the aggressive, expansionist nature of many agricultural economies, these techniques have also been applied to wild animals and to the natural world in general.

In his book *The Sacred Executioner*, Hyam Maccoby has referred to what he calls 'distancing devices'; that is, devices that distance the individual from the morally dubious consequences of his own actions.[2] Maccoby uses the concept in a lengthy discussion of human sacrifice, but it is equally applicable to the subject of animal exploitation and slaughter. Distancing devices of one sort or another permeate every aspect of our relationships with other species but, for the purposes of the present discussion, they can be loosely assigned to four main categories: *detachment, concealment, misrepresentation* and *shifting the blame*. Of the four, detachment is probably the most widespread.

DETACHMENT

A substantial number of people in modern, western society grow up with scarcely any contact with animals or nature.

Those raised within cities may not even be familiar with pets, and their experience of other species is often limited to negative encounters with houseflies, cockroaches, rats, mice, sparrows, starlings, pigeons and other urban 'pests'. It is easy for such persons to feel detached from or indifferent to animals, since animals have never played an important role in their lives. In more traditional rural settings, or, indeed, in any setting where humans and animals are obliged to live in close contact, detachment generally requires a certain amount of conscious or deliberate effort to maintain. In the opening paragraph of *Man Meets Dog*, the ethologist, Konrad Lorenz, illustrated the nature of the problem:

To-day for breakfast I ate some fried bread and sausage. Both the sausage and the lard that the bread was fried in came from a pig that I used to know as a dear little piglet. Once that stage was over, to save my conscience from conflict, I meticulously avoided any further acquaintance with that pig.[3]

Lorenz was generally regarded as a gifted naturalist with an acute understanding of animal psychology. This insight won him the Nobel Prize, but clearly it also generated problems. His ability to empathize with the pig and to perceive it as something intrinsically endearing raised ethical consider-ations that threatened to spoil his breakfast. So the simple solution, from Lorenz's point of view, was to avoid getting to know the animal in the first place in order, as he says, to save his conscience from conflict. Whenever human beings are obliged to maltreat or kill animals, they display a strong tendency to detach themselves in this way. As Miriam Rothschild points out, 'just as we have to depersonalize human opponents in wartime in order to kill them with indifference, so we have to create a void between ourselves and the animals on which we inflict pain and misery for profit'.[4] Detachment can also be learned. People who, for whatever reason, kill or harm animals regularly and frequently, tend to become progressively desensitized to the experience. The first few episodes are generally disquieting, but after a period of time, the person habituates and the act becomes a reflex, virtually devoid of emotional content.* Such people may take pride in the technical skills needed to perform their work efficiently, but it is generally a detached sort of pride in which the animal is viewed primarily as an artifact.[5]

On the whole, the amount of effort needed to maintain detachment increases the closer – in every sense of the word – the animal is to people. Thus, it is psychologically easier to shoot a rabbit at a distance than to club it senseless with a stick, and people generally have fewer inhibitions about killing fish than killing warm-blooded birds or mammals that more closely resemble human beings. The majority of humans will swat a fly or crush an ant to death with the same air of detachment that they might remove a speck of dirt from their clothes. Aristotle's concept of a natural hierarchy, and Descartes's view of animals as machines were both myths that aided the process of detachment by creating an absolute rather than a relative distinction between humans and non-humans.

In traditional systems of livestock husbandry, it is difficult, if not impossible, for the stockman or herder to detach himself from his animals. To do so, as we have seen, would be self-defeating. Indeed, predominantly pastoral peoples, such as the Nuer of Sudan, develop remarkably close bonds with their animals. According to the anthropologist, Evans-Pritchard:

> The Nuer and his herd form a corporate community with solidarity of interests, to serve which the lives of both are adjusted, and their symbiotic relationship is one of close physical contact. The cattle are docile and readily respond to human care and guidance. No high barriers of culture divide men from beasts in their common home . . .

> Cattle are the Nuer's most cherished possessions. They are the focus of social and religious life. Every man knows the

---

* There are, of course, exceptions to this rule. Arluke (1994), for example, provides a thoughtful analysis of the more complicated, ethical coping strategies employed by animal shelter workers who are responsible for the routine destruction of other people's unwanted pets.

habits of his cattle; their individual idiosyncrasies; which are restless, which are troublesome, which like to drink on their way to pasture, and so forth. When a young man's ox comes home in the evening

he pets it, rubs ashes on its back, removes ticks from its belly and scrotum, and picks adherent dung from its anus. He tethers it in front of his windscreen [sic] so that he can see it if he wakes, for no sight so fills a Nuer with contentment and pride as his oxen. The more he can display the happier he is, and to make them more attractive he decorates their horns with long tassels, which he can admire as they toss their heads and shake them on their return to camp.

Predictably, the intimacy of this relationship, its lack of detachment, imposes restrictions on the slaughter of cattle. Cattle are valued by the Nuer primarily as a source of milk which, together with sorghum, is the staple food. The Nuer also bleed their cows and consume the blood in various forms, but they deny that they perform the bleeding operation in order to obtain food. Rather it is justified on the grounds that periodic bleeding is beneficial to the animal's health. Although the Nuer enjoy eating meat, it is considered bad form to slaughter a beast solely to satisfy hunger. It is even believed that the animal may curse them if they do so. Any animal that dies of natural causes can be eaten but, otherwise, cattle or, indeed, sheep and goats are only slaughtered at special ceremonies, ostensibly to honour or propitiate ghosts and spirits.[6] These ceremonies of propitiation are entirely different from those practised by typical hunters, although the underlying motive is the same. The hunter does it in order to atone for the animal's death. The Nuer kills the animal because, as he sees it, the spirit world demands a sacrificial gift as a placatory token of good faith. In other words, the Nuer appears to *shift the blame* for the killing from himself to some supernatural entity, so that the meat can be eaten with a clear conscience.

An almost identical relationship exists between the Tungus of central Siberia and their domestic reindeer. Here the animals are kept chiefly as a source of milk and for riding, and once again they are only slaughtered in order to appease capricious and potentially malevolent supernatural powers. Like the Nuer, the Tungus are very fond of their livestock. They display a deep affection for them, caress them, talk to them and believe that the animal understands what is said. They know each of them individually by a name to which the reindeer learns to respond, and go to considerable lengths to protect them from swarms of mosquitoes and predatory wolves. In his book, *Hunters, Pastoralists and Ranchers*, Tim Ingold denies that the Tungus' unwillingness to slaughter their reindeer has anything to do with their close familiarity or identification with people. He prefers a strictly pragmatic, economic interpretation based on the fact that the animals are more valuable alive than dead. Yet, in the following paragraph, he asserts that the Tungus prohibit the spilling of blood during a sacrifice because a bloody slaughter 'within the domestic domain would, presumably, be tantamount to murder'.[7]

Ingold also notes that the shift from the more traditional domestication practised by the Tungus to the more intensive reindeer ranching adopted by other Siberian peoples has been accompanied by increasing signs of detachment. He mentions the 'markedly secular attitude on the part of ranchers toward animals and their treatment' and observes that the ritual respect normally accorded to reindeer is entirely lacking in this setting. Detachment is reinforced by an increasing level of violence directed at the animals themselves. Whereas the Tungus perceive their reindeer as quasi-human subjects who willingly submit to human contact and control, the ranchers view their beasts as intractable objects which are hellbent on escape, and which therefore need to be forcibly subdued and coerced.[8] The gratuitous violence involved in this reinforces detachment by making the animals fearful of humans and therefore less inclined to act socially toward them.

Detachment and unnecessary brutality seem to be universal components of intensive animal husbandry, presumably because they help to distance the farmer from the mass suffering and slaughter for which he is either directly or indirectly responsible. The process is also encouraged by the advent of modern farming methods. Technological advances within the past century have brought about a steadily increasing physical separation between livestock and their keepers. In the most intensive systems, much of the day-to-day care of the animals is performed by machines, and their daily food intake is controlled and monitored by computer. Even their housing appears specially designed to minimize human contact. If he likes, the owner – the person who decides the destiny of the animals – can avoid contact with them altogether. Even if his farm is not extensively mechanized, the chances are he will employ someone else to do the dirty work, someone who will have the opportunity to get to know the animals, but who will not be held personally responsible for their fate. At this level of detachment, the animal easily becomes a mere cipher, a unit of production, abstracted out of existence in the pursuit of higher yields.

But higher yields and maximum productivity are only part of the story. Time and again, animal welfare organizations and committees have suggested sensible, humane alternatives to the more intensive forms of factory farming. Some of these suggestions could be implemented without substantially increasing expenditure, and a few might even reduce overall costs by trimming overheads.[9] These same groups have also conducted surveys the results of which indicate that a majority of consumers would be prepared to pay slightly more for animal products produced by less intensive methods.[10] Yet far from provoking a spate of humane legislation and reform, as would seem morally appropriate (not to say imperative), such findings appear to have made the farming community more entrenched than ever. This kind of resistance seems to have little to do with straightforward economics, but it can be easily understood in terms of the conflict between sympathy and exploitation. Not surprisingly, modern livestock producers are generally unwilling to consider the welfare of their animals because this would entail thinking about them as subjects rather than objects; as persons rather than things, and this would raise imponderable questions about the morality of their treatment. Far easier and less painful to treat them as Cartesian automata that can be slaughtered and exploited according to the dictates of market forces and nothing more. Of course, there are many farmers to whom these remarks do not apply, and many others who justifiably *shift the blame* to the consumer. Nevertheless, detachment as a distancing device is an important ingredient of intensive animal exploitation, and one of the most serious impediments to humane reform.

When necessary, the human capacity for ruthless detachment can be remarkably selective. The native inhabitants of Polynesia, for example, kept dogs as pets and demonstrated a deep affection for them. But they also raised dogs commercially in order to serve as items of food. In 1866, George Jesse professed himself puzzled and disturbed by this curious double standard:

It is strange that this gentle and manly race of beings should not have had their sympathies more entwined with the creatures so much partaking of their character. That mothers should suckle an animal, and yet allow that same race to become an article of food, is a singular contradiction in feeling.[11]

Yet this seemingly contradictory behaviour makes more sense when one discovers that the dogs that were kept as companions were rarely slaughtered or eaten, and then only under severe protest from the owner, while the animals raised for slaughter were never, as far as it is possible to judge, befriended.[12] In other words, the Polynesians remained emotionally and physically *detached* from the dogs that were destined for the pot, but allowed themselves to become exceptionally *attached* to particular pet dogs which were then exempt from slaughter.[13]

A similar dichotomy of attitudes can also be found in certain Eskimo or Inuit societies. Judging from reports, a dog's life among the Inuit is far from pleasant. One observer noted that he 'never heard a kind accent from the Eskimaux to his team', and that:

... the driver's whip of walrus hide, some twenty feet long; a stone or a lump of ice skillfully directed; an imprecation loud and sharp, made emphatic by a fist or foot; and a grudged ration of seal's meat, make up the winter's entertainment of an Eskimaux team.[14]

Individual Inuit will express pride in the strength or endurance of their dogs but ordinarily they will avoid physical contact with them and show them little overt affection. But there are interesting exceptions to this rule. Some Inuit, especially childless couples, will adopt a puppy and lavish affection on it. Thereafter, these pets are not expected to engage in any kind of useful work.[15] Again, all this makes sense. Dogs have no natural inclination to pull sledges, as they have for, say, hunting or providing companionship. They need a lot of encouragement to do so, and in bad weather, or when hungry and exhausted, they need to be physically coerced. Of course, it is possible to train dogs to pull sledges using rewards rather than punishments, but the average Inuit is obliged to dispense with such niceties. During the arctic winter, his life may depend on getting from A to B in the shortest possible time and he can hardly be expected, under the circumstances, to worry himself about canine welfare. So he curses and kicks his intransigent dogs in the same way that you or I might curse and kick a car that broke down on the way to an important engagement. The Inuit's relationship with his sledge-dog is therefore essentially that of driver and slave. Within it, sympathy and affection are misplaced and potentially hazardous, unless the animal is first classified separately as a companion rather than as a beast of burden.

Examples of similar selectivity are common in our own society. Laboratory researchers who use animals in experiments, for example, will sometimes single out one or two individuals and keep them as pets or mascots. They may develop strong attachments for these 'pets' and generally will not consider using them for research, although they will happily conduct experiments on their more anonymous counterparts.[16] In an article entitled 'The Moral Status of Mice' the psychologist, Harold Herzog, provides a particularly good example of this human capacity to arbitrarily award or deny moral standing to different classes of animals. Herzog identifies three different classes of mice in American research laboratories which he labels 'good mice', 'bad mice' and 'feeders'. Good mice are the research animals who give their lives for the benefit of humans. Their care and husbandry is regulated by official animal welfare guidelines, and their experimental uses are carefully scrutinized and vetted by Institutional Animal Care and Use Committees. Bad mice are essentially good mice that have escaped from their cages and gone feral. As soon as these mice hit the floor, so to speak, they experience an immediate loss of moral status. It is permissible to use any means to exterminate them, including techniques which cause severe and protracted suffering – techniques, moreover, that would not be permitted if they were part of a scientific experiment. Finally, there are the feeder mice which are bred and raised purely to serve as food for other laboratory animals, such as reptiles. As long as these mice are used for routine feeding, no regulations apply. But if the feeding is part of a scientific study of, say, predatory behaviour in snakes, the 'experiment' must first be approved by the institutional oversight committee concerned with animal welfare. The point which Herzog makes is that all of these mice belong to the same species, and may even belong to the same genetic strain. Yet it is still possible for people to treat them with widely varying degrees of ethical detachment merely by assigning them different roles or labels.[17]

One could likewise argue that the western practice of lavishing affection on dogs and cats, while turning a blind eye to the callous exploitation of pigs, poultry and veal calves, is yet another case of precisely this kind of selective detachment.

CONCEALMENT

On the principle that what the eye does not see the heart does not grieve over, *concealment* is the natural partner of detachment. On more traditional farms, effective concealment of animals and animal suffering is obviously difficult to secure. In modern, intensive systems it is relatively easy and widespread.

Factory-farmed pigs and poultry are kept in anonymous-looking, windowless buildings that more closely resemble warehouses than animal enclosures. Once inside, the animals are out of sight and, effectively, out of mind as far as the majority of people are concerned. Their transportation to the slaughterhouses also tends to be performed surreptitiously, and the abattoirs themselves tend to be hidden from public view. Since the Middle Ages, civic authorities have done their best to prohibit the slaughter of animals in public, and nowadays most slaughterhouses are situated away from residential areas on the outskirts of towns and cities.[18] Many meat-eating consumers react with horror to the sight of a recently butchered carcass, and it is clear that people do not like to be reminded that the plucked and trussed-up chicken, or the leg of lamb they had for Sunday lunch was once a warm-blooded, sentient life-form like themselves.

There is also concealment in numbers. On a small-holding with a milch cow, a couple of pigs, a small flock of sheep, and a dozen or so fowls, the farmer and his family are likely to get to know the animals individually, whether they wish to or not. But in a large-scale production unit containing thousands of superficially similar animals crammed into identical pens or cages, individual identification becomes so difficult that animals have to be numbered, tatooed, freeze-branded, tagged or ear-notched for the purpose. Any sort of personal rapport with an animal in these conditions is well-nigh impossible and, although few would care to admit it, this is probably intentional.

Verbal concealment is also commonplace. We talk about 'beef', 'veal' and 'pork' rather than bull-meat, calf-meat or pig-meat because the euphemisms, in every sense, are more palatable than the reality.[19] The meat industry is only too well aware of this. A recent edition of the British *Meat Trades Journal* recommended a change in terminology designed to 'conjure up an image of meat divorced from the act of slaughter'. Suggestions included getting rid of the words 'butcher' and 'slaughterhouse' and replacing them with the American euphemisms 'meat plant' and 'meat factory'.[20]

Again, the need for verbal concealment decreases the further the animal is from a human being. Hence, we call chicken 'chicken', fish 'fish', and frog's legs 'frog's legs' without apparently spoiling our appetites. A recent controversy in Israel surrounded the sale of so-called 'white steak'. White steak is a euphemism for pork, a meat which is forbidden under Jewish dietary laws. In order to encourage sales, shop-keepers and restaurateurs invented the new term in order to conceal the true nature of the substance that was being eaten. It was not as if the people who consumed white steak were in any doubt about where it came from. It was simply easier for them to buy or order it than to use the taboo word 'pork'.[21] Verbal concealment crops up in most areas of animal exploitation. We speak of the 'harvest of the seas' as if fish and shellfish were analogous to wheat and barley, and in the fur trade 'pelts' are 'harvested' rather than animals flayed. It is even tempting to suggest that much of the technical jargon employed by scientists who experiment on animals is simply an elaborate euphemism, a method of disguising the animal's affinity with humans, and so promote detachment.[22]

MISREPRESENTATION

Another popular method of justifying the harmful exploi-tation of other creatures is by *misrepresentation*; by deliberately or unconsciously distorting the facts about them so that their suffering and death seems appropriate, necessary or deserved.* The tendency to anthropomorphize animals does not invariably lead to sympathy and affection. It is just as easy for people to focus on an animal's unpleasant or dangerous human qualities as on its pleasant ones, and these apparently negative attributes can then be used as an excuse for killing it, brutalizing it or being indifferent to its welfare. In this

---

* The use of the term 'misrepresentation' in this context implies that there is such a thing as a *correct* or morally neutral way of representing animals. In reality, it is debatable whether humans are ever capable of viewing animals impartially, although, clearly, some representations are more subjective or value-laden than others.

context, the animal often becomes a projection of our own darker motives and desires, a symbol of the beast that lurks menacingly in the thickets of the human psyche.[23]

This sort of negative personification is particularly prevalent in human attitudes to predators, pests and scavengers – species that either compete with us directly, or which survive off the surpluses of human culture. Highly emotive, anthropomorphic language is often used to describe such species: they are spoken of as filthy, disreputable, gluttonous, sly, ruthless, evil, cowardly, blood-thirsty and savage. They are portrayed as dangerous and despicable enemies of society that invite nothing but hatred and loathing. Needless to say, these grossly inaccurate epithets serve nicely to justify their callous extermination.

In his book *Of Wolves and Men*, Barry Lopez has eloquently explored the countless myths and inaccuracies that have provided a pretext for annihilating wolves throughout most of the Northern Hemisphere. European and North American folklore has invariably and erroneously depicted the wolf as a malevolent, merciless and insatiable killer; the bane of winsome little girls in red riding-hoods; the archetypal *bête noire*. According to Lopez, the wolf was seen as the Devil incarnate: 'red tongued, sulphur breathed, and yellow eyed; he was the werewolf, human cannibal; he was the lust, greed, and violence that men saw in themselves'.[24] With such a reputation, however undeserved, it is scarcely surprising that wolves are now an endangered species throughout much of their original range.

Of course, the extermination of wolves and the mendacious propaganda that encouraged their persecution had a sound economic purpose. As civilization expanded, as the forests were cut down and the wildernesses cultivated, the wolf's natural prey gradually disappeared, and he was forced to turn his predatory attentions to domestic livestock. In this new role he undoubtedly posed a serious threat to human livelihood. But the violence and hostility which his depredations aroused went far beyond the bounds of economic necessity. Wolves were not just killed; they were tortured, set on fire, poisoned, mutilated and hunted down with pathological dedication. And the process, unfortunately, still continues. As recently as the 1970s the inhabitants of Minnesota choked timber wolves to death in snares to show their contempt for the animal's designation as an endangered species.[25] In 1982, wolf fever gripped Norway when a single animal – virtually the only one for miles around – was accused of savaging a flock of sheep. The Norwegian Fish and Wildlife Service promptly placed a massive bounty on its head, and sparked off a national panic resembling the hysteria created by the film *Jaws*. When the offending wolf was eventually cornered and shot in 1984 it became the object of extraordinarily tasteless festivities. The leader of the hunt, improbably named Lars Saga, poured champagne over it in his own sitting room for the benefit of eager journalists; he then drove the corpse to the local school and old people's home and, finally, to Oslo where he paraded it on the steps of the Norwegian Parliament.[26]

This level of unreasoning hatred and ferocity seems all the more bizarre and ironical when one remembers that the wolf is the wild ancestor of man's best friend, the domestic dog; an animal which, at least in western countries, is perceived as a paragon of self-sacrificing loyalty and devotion. Yet, in a curious way, this may be precisely the reason why it was necessary to fabricate such an outrageously distorted image of wolves. Those who have taken the trouble to study wolves in their natural setting have generally been struck by how likeable and essentially dog-like they are.[27] Such a conception would be incompatible with their ruthless extermination, so the real wolf had to be disguised under the cloak of the fiendish creature of legend.

An important element of this war against wolves and other 'vermin' was the concept of retribution. By personifying them as evil humans, we elevated them to the status of conscious, thinking beings who could be held personally responsible for their actions. Seen in this light, their acts of destruction were the acts of criminals, individuals who deliberately conspired to cause others harm. According to law, criminals deserved retribution. They needed to be punished, both to set an example to others, and to restore the moral equilibrium of the community. Animals were subject to the same laws, and dangerous or noxious ones were therefore obliged to suffer the same penalties as human thieves and murderers. In Europe during the Middle Ages and the Renaissance, animals were regularly tried for a variety of crimes and executed, excommunicated or banished according to the circumstances. In 1457, for example, a sow and her six piglets were tried at Savigny-sur-Etang in France for murdering and partly devouring an infant. The sow was duly condemned and sentenced to death by hanging, but the piglets were pardoned because of their youth and innocence, the fact that their mother had set them a bad example, and because proof of their complicity was not forthcoming. Ecclesiastical cases were also brought against weevils, rats and grasshoppers for ravaging crops, and there was even a case of a cock that was burned at the stake in Basel, Switzerland, because it layed eggs. Animals were also imprisoned and tortured before execution, not with a view to extracting a confession, but merely out of a slavish desire to carry out the letter of the law.

Such accounts may seem farcical but they were based on one of the most ancient codes of justice: the *lex talionis*, or the principle of an eye for an eye, a tooth for a tooth. Under this code, someone or something had to pay the penalty for every crime, or else the community itself would suffer divine retribution in the form of famine, disease and general misfortune.[28] The idea of applying this rule to birds and beasts may appear quaint, but it is worth remembering that animals are still misrepresented in similar ways. Many pet-owners firmly believe that their animals 'know' when they have done something wrong, and punish them for transgressing human rules of conduct. Likewise, on British shooting estates, it is still relatively common to see gamekeepers' 'gibbets' where the corpses of weasels, stoats, rats, crows and other vermin are displayed, ostensibly to serve as a warning to others.

One of the reasons why the misrepresentation of animals is so widespread and difficult to counteract is that it is frequently self-reinforcing. In many parts of the world, for example, dogs are regarded as unclean and despicable creatures because they are seen to engage unashamedly in activities that are prohibited or taboo in human society; activities such as sexual promiscuity, incest, indolence and the eating of carrion and faeces. In other respects, dogs are very human-like and easy to identify with, so they readily serve as potent symbols of human moral degeneration and depravity. Among the Maoris of New Zealand it was even believed that dogs originated from a human being who was transformed into a dog as a punishment for social misbehaviour.[29] The net effect of defaming dogs in this way is that they become outcasts or pariahs, forced to eke out a living on the lowest fringes of human society. But as the saying goes 'give a dog a bad name'. By casting them out, by obliging them to fend for themselves, societies like these more or less force the animal to perform the very acts for which it is despised.[30] The Arab who shuns the dog, who drives it away with imprecations and curses, and who makes it live in the streets off carrion and ordure, simply reinforces his own loathing. Similarly, the westerner who claims he dislikes dogs because of their fawning servility, tends to behave toward dogs in a way that is almost guaranteed to provoke further demonstrations of ingratiating friendliness. The important thing to remember in both cases is that neither representation of the dog is in any sense accurate. Both are projections of characteristics people fear and despise in themselves, and both serve to justify the avoidance or ill-treatment of the animal in question.

Something very similar to this exists in our treatment of the domestic pig. The word 'pig' is a widespread term of abuse. When we call someone a pig or refer to their behaviour as swinish we focus on certain qualities of the animal, such as the way it eats or its fondness for wallowing in mud, and we then evaluate these behaviour patterns as if they were being performed by a person. In other words the pig, like the dog, has become a metaphor for a gross and degraded human lacking in social graces. Unfortunately, the image is horribly reinforced by the way we keep pigs. We compel them to live in overcrowded, overheated, foul-smelling piggeries. We

condition them to engage in manic feeding frenzies by
controlling their food supply, and we force them to stand
and sleep on concrete or metal floors awash with their own
excrement (see chapter 1). The mere fact that they manage to
survive these conditions, let alone remain outwardly healthy,
simply helps to confirm the suspicion that they are, indeed,
debased and insensitive creatures unworthy of moral concern.

It is not invariably necessary for humans to misrepresent
animals disparagingly in order to justify harming them. In the
rhetoric of modern sport hunting, for example, the quarry is
often affectionately depicted in admirable or praiseworthy
terms. Its status is elevated to that of a worthy opponent in
an amusing game of life and death; an opponent who enjoys
nothing more than pitting his own strength, speed or cunning
against that of a well-armed human.[31] By weighting the
imaginary odds against themselves in this way, sport hunters
seem to be able to kill with a clear conscience while simultaneously affirming their own superiority over the vanquished
prey. They may also derive feelings of enhanced social and
moral prestige relative to other hunters who pursue less
formidable animals that are easier to kill.

SHIFTING THE BLAME

Once a year at the great Bouphonia or bull-slaying feast in
Athens, a bull symbolizing Zeus, the father of the gods, used
to be sacrificed on the temple altar. Immediately after the
sacrifice, the priests who were responsible for the killing fled
from the altar in mock panic, crying out a formula that
absolved them of guilt. Later a trial was held in which the
blame for the slaying was attributed to the sacrificial knife
which, having been found guilty, was then punished by being
destroyed.[32]

This ritual – in which, incidentally, the bull's carcass was
eventually devoured by the populace – provides a clear
illustration of the human tendency to evade responsibility for
morally culpable acts by *shifting the blame*. In this particular
case, a knife is held responsible and is punished accordingly,

but in most other societies the guilt most often comes to rest on other animals, persons or supernatural powers. Various examples of blame shifting have already been mentioned. Siberian bear-hunters blame 'the Russians', the Chenchu blame their dogs, and the Moi of Indochina blame the animal itself for past misdeeds. Similarly, in our own society, destroyers of so-called vermin perceive their victims as reprehensible miscreants who deserve death. As with most distancing devices, blame shifting becomes increasingly formalized and contrived when associated with the slaughter or maltreatment of tame or domestic animals.

In the highlands of New Guinea, pigs are probably the most important domestic animals. Traditionally, the care of pigs is the exclusive prerogative of women who adopt these animals and keep them as pets. Piglets are suckled at the breast, and carried around like human infants. When they are older they are hand-fed, petted, groomed, decorated and generally made much of. Not surprisingly, New Guinea women develop strong emotional attachments for these animals, and the pigs in turn become as sociable and affectionate toward their foster mothers as dogs. Yet periodically, these same people hold prodigious pig feasts in which vast numbers of their porcine companions are slaughtered and consumed. This seemingly contradictory state of affairs is made possible by shifting the blame. In the first place, the women do not knowingly eat their own pigs. The primary function of the pig feasts is socio-political; they are designed to curry favour with neighbouring groups or clans who are invited over to indulge in an orgy of pork-eating. Secondly, while the women care for and nurture the pigs, it is invariably the men who take them away and slaughter them. Thanks to this relatively simple division of sex-roles, the women are able to rear the pigs with all the necessary affection and sympathy, and yet remain largely blameless for their destruction. The men, meanwhile, can slaughter the pigs with apparent detachment because they have had little or nothing to do with the animals personally. It is perhaps worth mentioning in passing that the relationship between New Guinea men and women is distinctly strained at the best of times. Husbands and wives live apart in separate houses, and men believe that too much contact with women can cause everything from premature baldness to permanent brain damage.[33]

In our own society, division of labour within the animal farming and animal marketing community seems to serve much the same purpose. One farmer may specialize in breeding pigs and producing piglets, another grows the piglets on to slaughtering weight, another person transports them to the abattoir and another is responsible for killing and butchering them. The buck or, in this case, the pig is passed from one individual to another down the line before landing fairly and squarely in the lap of the consumer. In a sense, everyone involved is guilty, but no one is obliged to shoulder the full burden of responsibility.

In former times, when there were fewer middlemen involved in the process, it was the butchers and slaughtermen who carried most of the blame. Throughout history, those directly responsible for killing animals have been regarded with a curious mixture of awe and disgust, not unlike that normally reserved for public executioners. In England, they have had a dubious reputation since at least the Middle Ages and, in seventeenth-century literature, were frequently described as odious, merciless, pitiless, cruel, rude, grim, stern, bloody and greasy. In 1716, the poet, John Gay, advised Londoners:

> To shun the surly butcher's greasy tray.
> Butchers, whose hands are dyed with blood's foul stain,
> And always foremost in the hangman's train.

Likewise, in 1748, the philosopher David Hartley referred to the 'frequent hard-heartedness and cruelty found amongst those persons whose occupations engaged them in destroying animal life'. Whether or not there was any substance to this prejudice is difficult to ascertain, yet the notion that these people were inherently callous and cruel was so prevalent that many considered them ineligible for jury service in capital cases.[34] The slaughterman's profession is still viewed with distaste by many people. A recent pro-vegetarian polemic, for example, states that it is certainly true that the slaughterer's occupation is:

grim and brutalizing. Few people work in stockyards by choice. Most are there because their families have worked in the business; many are illegal immigrants. Workers are forced to become indifferent to the vocal protests and struggling of the animals they kill. It is likely that the callousness they develop in order to endure the realities of their jobs will affect other areas of their lives.[35]

In the great civilizations of antiquity, slaughtermen and butchers also had unusual status. They were almost invariably priests. The association between the ancient religions and animal sacrifice has been the subject of a great many scholarly works, and an almost equal number of theories and hypotheses. The general consensus seems to be that the ancients considered animal sacrifice necessary as a means of satisfying the appetites of the various supernatural powers who were believed to govern the destiny of human beings. These powers were conceived of in highly anthropomorphic terms. They were temperamental and easily provoked, but they could also be placated by means of appropriate gifts and offerings. Like humans, they needed feeding and, like humans, they regarded meat – or at least the souls of meat-bearing animals – as the most nourishing and appetizing food. Failure to keep the gods satiated was guaranteed to inspire their malice, and when the gods were offended, tempests, droughts, pestilence and famine inevitably ensued.[36] The ancient Egyptians, Mesopotamians, Hebrews, Greeks, Romans and Aztecs had every reason to worry about the forces of nature. The agricultural and animal-based economies on which they depended were undoubtedly productive and supported much larger human populations than those of subsistence hunters. But they were also notoriously vulnerable to sudden environmental changes. A freak storm or flood, a shortage of rain at a crucial time, or unpredictable outbreaks of pests and diseases could obliterate an entire year's harvest or wipe out whole herds of livestock. The system was inherently precarious, and the higher population density made the ensuing epidemics and famines all the more horrific. The uncertainty of it all generated intense anxiety, and these fears were personified in the form of spiteful and capricious gods that had to be appeased. Occasionally, fears were so exaggerated that the ultimate sacrifice was deemed necessary. Animal victims were then replaced with humans.[37]

Institutionalized animal sacrifice also served another purpose. It provided a useful method of shifting the blame. In most of these civilizations, any slaughter of domestic animals was regarded as a sacrifice that could only be performed by those properly versed in the sacred mysteries. In other words, by the priesthood. Temples were not only places of worship, they were also abattoirs and butcher shops all rolled into one. In his book, *Homo necans*, Walter Burkert describes a typical ancient Greek sacrifice in some detail. First, participants in the ritual symbolically purified themselves by bathing and putting on clean clothes. Sexual abstinence was also recommended. The sacrificial animal was also cleaned, groomed and decorated. It was generally hoped that the beast would follow the sacrificial procession voluntarily, and legends often told of animals that willingly offered themselves for slaughter. On arriving at the altar, the priests again washed themselves. Water was also sprinkled on the animal's head, while it was encouraged verbally to shake itself. The nodding of the head was then taken as a signal of assent, a sign that the beast concurred with its own slaughter. As in the Ainu bear sacrifice, the victim was also mildly taunted by having objects thrown at it, usually harmless things such as barley grains, but occasionally stones were also hurled. In some cases, the animal was also rendered culpable by being 'allowed' to eat sacred cakes which were left on the altar for this purpose. The executioner then moved toward the animal with the sacrificial knife carefully concealed from sight. As the death blow was struck, and the animal's blood flowed out over the altar, the women signalled the emotional climax of the event by uttering long, wailing screams. The sacred act completed, the carcass was carved up and disembowelled. The heart, sometimes still beating, was placed on the altar, and the lobes of the liver carefully scrutinized for auguries of the future. The rest of the edible remains were then roasted on the altar fire and promptly devoured by everyone present. Only the bones and small selected morsels of flesh were burned entirely as an offering to the gods.[38]

Burkert describes animal sacrifice as 'an all-pervasive reality in the ancient world', and it appears that many of the ancient Mediterranean cultures considered the consumption of unsacrificed meat taboo.[39] Although the details varied from place to place, the same basic ritual elements characterized Egyptian, Phoenician, Babylonian, Hebrew, Persian, Etruscan and Roman sacrificial customs. In every case, the performance constituted an elaborate exercise in blame-shifting. The animal was delivered to the temple or priest, it was made to appear willing or acquiescent, and the priest then performed the slaughter. A token amount of meat was then burnt as an offering to the deity, and the remainder was either returned to the owner – after payment in kind – or redistributed or sold to the general population. The priests were directly responsible for the animal's death, but theirs was a sacred duty and therefore forgivable. Besides which, they took elaborate ritual precautions in order to cleanse themselves of guilt. It was ultimately the gods who were to blame, since it was they who demanded the sacrifice in the first place. This is not merely speculation. According to an ancient Babylonian text, the head priest actually bent down to the ear of the slaughtered victim and whispered, 'this deed was done by all the gods; I did not do it.'

As Burkert points out, the god to whom these ancient sacrifices were made seemed little more than 'a transparent excuse for festive feasting'.[40] And many early classical scholars considered the rite objectionable precisely because the underlying carnivorous motives were so obvious. In Greece, the Pythagoreans and Orphics, who believed in metempsychosis, extolled the virtues of vegetarian sacrifices, such as barley cakes, honey or oil, while Empedocles (495–435 BC) condemned sacrifice as a form of cannibalism since, in his view, animals were simply reincarnated humans. Theophrastus, Aristotle's pupil and successor, argued that animals, by their nature, share kinship with us, and that it is therefore unjust to rob them of life. Both he and Empedocles, regarded sacrifice as a shameful and morally polluting activity, as well as being a corrupt deviation from earlier customs and practices. Plutarch (AD 46–120) reasoned that humans were not naturally carnivorous since we lack the appropriate organs of predators and eat our meat cooked. He challenged his critics to catch animals with their teeth and devour them raw. Finally, Porphyry (AD 232–309) synthesized all of these ideas together in his treatise *On Abstinence from Animal Food* – a thorough and penetrating criticism of both sacrifice and meat-eating. In his view, the only worthwhile offering to be made to the gods was 'a pure intellect and an untroubled soul'.[41]

Biblical references to sacrifice also express a degree of ambivalence. In Psalms (51: 16 & 17), for example, we hear that the Lord 'desirest not sacrifice' and 'delightest not in burnt offering'. Rather 'the sacrifices of God are a broken spirit: a broken and a contrite heart'. The prophet Isaiah also rages against the excessive use of animal sacrifice:

To what purpose is the multitude of your sacrifices unto me? saith the Lord: I am full of the burnt offerings of rams, and the fat of fed beasts; and I delight not in the blood of bullocks, or of lambs, or of he goats. (Isa. 1: 11)

Nevertheless, during the post-exilic period, animal sacrifice appears to have been a central feature of the Jewish religion. Bullocks, rams, goats, lambs, chickens, turtle doves and pigeons were regularly sacrificed by the Israelites, and the smell of burnt offerings was deemed to be a 'sweet savour unto the Lord' (Lev. 2: 2). Despite being given God's sanction to eat the flesh of animals, Jews were allowed to do so only with the proviso that they return the animal's 'life' – in other words, its blood – to God: 'the life of all flesh is the blood thereof: whosoever eateth it shall be cut off' (Lev. 17: 14). According to the rules, the blood of hunted animals had to be poured out onto the ground and covered with dust, while domestic livestock had to be sacrificed on temple altars where their blood would constitute an offering to the Lord. Killing livestock without the appropriate priestly ritual was equivalent to murder: 'blood-guilt shall be imputed to that man; he has shed blood; and that man shall be cut off from among his people' (Lev. 17: 4). Classifying an animal's blood as its life, and then returning that life-blood to God, was clearly another way of deflecting blame; a subtle means of exonerating the butcher from blood-shed.[42]

Christians did not employ animal sacrifice, except in isolated instances.[43] Yet, significantly, the Christian Church was founded on the image of Christ, the Lamb of God, meekly offering himself for slaughter in order to atone for the sins of the world. To commemorate this supreme sacrifice, Christians even engage in a sacramental meal in which the body and blood of the Saviour are symbolically devoured.[44] Unfortunately, instead of promoting a more honest and responsible attitude to animal exploitation and slaughter, Christianity gave rise to the ultimate expression of blame-shifting with the idea of God-given human supremacy over the rest of creation. In the religions of antiquity, animals that were sacrificed were generally treated with respect, and were sometimes pampered and fêted for a year before dying under the sacrificial knife. Their deaths were seen as necessary but, at the same time, sufficiently culpable to warrant some form of preliminary recompense. Under Christianity, this sort of respect for the animal's feelings became entirely superfluous. Animals had only one purpose in life, and that was to serve human beings. They owed their existence to God, and God had ordained that they should suffer under the yoke of human dominion. If anyone was to blame, it was God, but of course God was Almighty and therefore beyond reproach.

The foregoing catalogue of distancing devices is very far from complete, and the reader will no doubt be able to think of many other examples of the same sort of thing. It is provided primarily to illustrate how the shift from traditional hunting to progressively more and more intensive systems of animal exploitation has been accompanied by the evolution of increasingly sophisticated methods of evading guilt. As a predatory species, it seems, we are confronted with a hideous moral dilemma. Our highly developed social awareness enables us to understand and empathize with animals, just as we understand and empathize with each other. It also allows us to use animals and to manipulate them to our own advantage in precisely the same way that it allows us, up to a point, to exploit and manipulate other humans. Yet this same empathic ability permits us to identify with animals personally, and to form close affectionate relationships with them which are themselves gratifying and potentially therapeutic. This is fine so long as the partnership is mutually rewarding, as is generally the case with human–pet relationships or the cooperative hunting alliance that exists between people and dogs. But it gives rise to unacceptable contradictions when our purpose in using animals involves their eventual slaughter, subjugation or maltreatment. As we have become increasingly involved with and dependent on domestic species, as we have pushed them further and further into the role of victims and slaves, the burden of guilt has grown to the point where it can no longer be expiated through simple acts of ritual atonement. So we have created an artificial distinction between us and them, and have constructed a defensive screen of lies, myths, distortions and evasions, the sole purpose of which has been to reconcile or nullify the conflict between economic self-interest, on the one hand, and sympathy and affection on the other.

Of course, the idea that people have inhibitions about killing and eating animals, such as pets, which are close to them emotionally is not new. It has been pointed out repeatedly in the past by a number of eminent anthropologists.[45] What has not been emphasized previously is the fact that close social bonds with animals are themselves emotionally fulfilling, and that they therefore constitute a benefit which frequently conflicts with economic demands. It is not so much that we avoid killing the animals with which we are friendly. It is more the other way around. Unconsciously or deliberately we either avoid befriending the animals we intend to harm, or we fabricate elaborate and often mythological justification for their suffering that absolves us of blame. The sad thing is that we have been practising this form of self-deception for so long that, by and large, we are scarcely aware that we are doing it any more. The myths have become reality, the fantasies, fact. Instead of questioning our supposedly objective, economic relations with other species, or the morality that governs our ruthless exploitation of animals and nature, we tend to ridicule or denigrate those who take the opposite view. People who display emotional concern for animal suffering, or the destruction of the environment, or the extinction of wild species are often treated as misguided idealists. While those who allow themselves to become emotionally involved with companion animals are considered perverted, pathetic or wasteful. And all of them are damned with the accusation of sentimentality, as if having sentiments or feelings for other species were a sign of weakness, intellectual flabbiness or mental disturbance. Yet, for more than 90 per cent of their history, human beings lived as hunters and gatherers, and the majority of hunter–gatherers display similar sentiments. The truth is that it is normal and natural for people to empathize and identify with other life forms, and to feel guilt and remorse about harming them. It is the essence of our humanity. The sooner we come to terms with this novel idea the better, since our future on this planet may depend on it.

CHAPTER 12

The fall from grace

Nature herself has, I fear, fastened on man a certain
instinct of inhumanity.

Michel de Montaigne, *Essays*

Judging from the biblical account of creation, Adam and Eve
were hunter–gatherers, or at least gatherers of wild foods. On
the day of his creation, the Lord is quite specific in saying to
Adam:

Behold, I have given you every herb bearing seed, which is upon the
face of all the earth, and every tree, in the which is the fruit of a tree
yielding seed; to you it shall be for meat. (Gen. 1: 29)

When the primordial couple made their fatal mistake and
gathered the fruit of the Tree of Knowledge, God's punishment was swift and unpleasant. Eve, the principal culprit, was
condemned to suffer labour pains, and to be perpetually
inferior to Adam. And Adam was forced to abandon his
carefree foraging existence and become a farmer:

Cursed is the ground for thy sake; in sorrow shalt thou eat of it all
the days of thy life; thorns also and thistles shall it bring forth
to thee; and thou shalt eat the herb of the field; in the sweat of
thy face shalt thou eat bread, till thou return unto the ground
. . . Therefore the Lord God sent him forth from the garden of
Eden, to till the ground from whence he was taken. (Gen. 3:
17–23)

It is interesting that the Book of Genesis depicts the change
from subsistence gathering to farming as a sort of punishment
– a decline in human living standards – since this is roughly in

line with the modern view proposed by the majority of anthropologists and archaeologists.

For centuries, westerners have cherished the belief that the development of agriculture and animal husbandry was a tremendous progressive leap forward along the road to economic prosperity. It was naturally assumed that the lives of Stone-Age hunters were short, brutish and uncomfortable – a perpetual struggle simply to find enough to eat – and that the invention of agriculture by some nameless Palaeolithic genius released our ancestors from their primitive shackles, and paved the way toward a better life, free from the toil and discomfort of subsistence hunting and foraging. This so-called Neolithic Revolution, it was supposed, allowed people to settle down in permanent villages, to produce surplus food, and to engage in all sorts of pleasant, unproductive activities, such as art, literature and science, upon which all the benefits of modern civilization were seen to rest. Recently, however, this reassuringly progressive, onwards-and-upwards view of human cultural development has been seriously questioned.

For one thing, existence at the end of the Palaeolithic appears to have been far more salubrious than previously supposed. Vast herds of grazing mammals – mammoth, bison, wild cattle and horses – roamed the fertile steppes and plains of Ice-Age Europe and North America, and provided Stone-Age hunters with a plentiful supply of food, protective clothing and other raw materials. Judging from the enormous accumulations of animal bones around their settlements, these people were also exceedingly adept at killing and butchering even the largest species, no doubt aided by a considerable armoury of beautifully fashioned stone, bone and wood implements and weapons.

Comparative studies of the remains of hunter–gatherers and subsequent agriculturalists indicate that the former suffered from fewer parasites and infectious diseases, and, generally speaking, enjoyed better health. Hunter–gatherers, for example, appeared to live longer than agriculturalists. The results of a large number of studies of archaeological populations from different parts of the world suggest that adult ages at time of death were higher on average in hunting and gathering groups than among early farmers. Rates of child and infant mortality also tended to be lower in pre-agricultural societies. Hunter–gatherers seem to have been better nourished as well. Roughly 30,000 years ago, at the peak of the glacial period, adult human males averaged 177 centimetres (5 ft 11 in.) in height, and females averaged 165 centimetres (5 ft 6 in.). Some 20,000 years later, when agriculture first began to appear, males had shrunk to the size of Ice-Age females, and women averaged no more than 153 centimetres (5 ft 0 in.). As far as it is possible to tell from their teeth, Ice-Age people were also in better condition. Thirty thousand years ago the average person died with only 2.2 teeth missing. In 6500 BC it had risen to 3.5, and by Roman times it was up to 6.6. Microscopic defects and developmental abnormalities in tooth enamel, thought to be indicative of early exposure to starvation and epidemic disease, also tend to be consistently more common in early agricultural populations compared with their hunting and gathering predecessors.[1]

Recent studies of living hunters and gatherers also indicate that this lifestyle is less arduous than that of most farmers. The anthropologist Richard Lee, for example, found that Bushmen living on the outskirts of the Kalahari Desert in Botswana spent less than three hours per day per adult collecting, hunting for or preparing their food. Yet, in return, they obtained a diet rich in protein and essential nutrients. The Bushmen spent the rest of their time resting, sitting around and chatting, entertaining visitors, doing handicrafts and visiting their neighbours.[2] This lifestyle compares very favourably with that of the average rural peasant who may spend some forty to fifty hours a week engaged in back-breaking farm work in return for a diet that is often deficient in protein.[3]

Of course, in certain select regions of the world, people nowadays enjoy a higher standard of living than ever before. But it would be a mistake to attribute this solely to the success of their agriculture. Past colonial exploits, and the present 'bubble' of industrialization, have made a substantial contribution to western material affluence, and the latter – because of its reliance on non-renewable mineral resources – may be expected to reach a point of diminishing returns in the not-too-distant future.[4] And even if we in the West have never had it so good, the global situation is virtually catastrophic. One estimate suggests that roughly a fifth of the world's population is permanently malnourished,[5] and there is little immediate prospect of any improvement. As one expert on the subject has observed:

Civilization has not been as successful in guaranteeing human well-being as we like to believe . . . Contemporary hunter–gatherers, although lean and occasionally hungry, enjoy levels of caloric intake that compare favorably with national averages for many major countries of the Third World and that are generally above those of the poor in the modern world. Even the poorest recorded hunter–gatherer group enjoys a caloric intake superior to that of impoverished contemporary urban populations. Prehistoric hunter–gatherers appear to have enjoyed richer environments and to have been better nourished than most subsequent populations (primitive and civilized alike).[6]

Given that the switch to farming seems to have been accompanied by an overall decline in human living standards, one wonders why Stone-Age hunters ever bothered to make the change. The probable answer is that they were forced into it whether they liked it or not.[7] In his book *Cannibals and Kings*, the anthropologist, Marvin Harris, has discussed the possible factors that induced this change of heart. As already stated, the last Ice Age was a period of relative plenty, a sort of Golden Age. Game animals were abundant, and Palaeolithic hunters had all the necessary technology and expertise to kill and butcher these animals efficiently. It is likely that human populations increased in size in response to this surplus of food. The relationship between these hunters and their prey would probably have reached some sort of stable equilibrium had conditions remained the same, but unfortunately they did not. Roughly 13,000 years ago, the climate suddenly began to get warmer, and the ice caps that covered much of the

Northern Hemisphere began to retreat, leaving behind them an ecological disaster area. As the vegetation altered from grassland to forest, the great herds of mammals underwent an immediate decline, and many became extinct, a process that was doubtless exacerbated by over-hunting.[8] People responded to the crisis in two ways. They intensified their foraging efforts and thus hastened the decline of game and other natural produce. And they diversified their diet to include a much wider range of less nutritious foods, particularly plants. Eventually, in certain areas, the situation became so bad that hunting and gathering on its own was no longer a profitable means of subsistence. At the time, people probably knew enough about plant biology to experiment with cultivation, but up until then it had not been worth the extra effort. But as suitable wild foods became gradually scarcer, farming became the only viable alternative.[9]

Although labour intensive, agriculture was an attractive proposition because, at least in the short term, it did produce an abundance of highly calorific food. But this, in turn, produced an increase in female fertility, allowed further population growth, and necessitated the adoption of more and more intensive systems of land use to feed the growing numbers of people.[10] An almost universal characteristic of farming is that it tends to be expansionist, and there is a very good reason why. When land is cleared and tilled, rainfall immediately begins to wash essential nutrients from the soil, while the removal of crops for human consumption creates a further drain by interrupting the normal nutrient recycling process. The gradual decline in soil fertility means that sustained use of the same piece of land is impossible unless the farmer has some way of replacing what he has removed.[11] In other words, he is under constant pressure to exploit formerly virgin, uncultivated land. The process of soil deterioration can be arrested or slowed down by techniques such as shifting cultivation, crop rotation or the regular application of compost, manure or artificial fertilizers, but only, it seems, temporarily. The history of agriculture is essentially one of continual expansion and encroachment into natural wilderness areas; the replacement of stable, complex biological communities with man-made systems dominated by a handful of domestic species.[12] The process, moreover, is rarely reversible. Land which has been over-cultivated or over-grazed takes decades or even centuries to regenerate. Hunting and gathering becomes increasingly uneconomical because the wild animals and plants on which this lifestyle depends are pushed aside and exterminated during the process of expansion. In short, the Neolithic farming revolution was effectively a one-way ticket. The beginning of a journey of no return.

As suggested in the previous two chapters, the shift from hunting to farming obliged human beings to adopt a radically different attitude to the natural world. The typical hunter takes what he needs for his own survival but, otherwise, he exerts a comparatively minor influence on the environment. His attitude to nature is predominantly one of respect or reverence rather than superiority. Guilt about killing animals or causing wanton environmental damage inhibits him from over-exploiting resources, and allows him to coexist with his food supply relatively harmoniously.[13] This respectful attitude to animals and nature has been described in a wide variety of extant hunter–gatherer societies, and some authors have even referred to the 'conservation ethic' typical of such groups.[14] For example, in his description of the Koyukon of central Alaska, the anthropologist, Nelson states that:

One of the most pervasive and potent themes that emerges in Koyukon ideology is a prohibition against waste of anything in nature. Strong sanctions apply to killing animals or plants and leaving them unused. Meat is carefully butchered and stored where it will not spoil . . . and fullest possible use is made of it to avoid offending the animal's spirit . . . Koyukon hunters go to great lengths to avoid losing wounded game. If a shot animal escapes, it is doggedly pursued, every effort is made to retrieve it, and if it is not found the hunter is genuinely upset. Most animal meat and organs are utilized, and disposal of the parts considered unusable is carried out in special, respectful ways . . . Among the Koyukon, reverence for nature, which is strongly manifested in both religion and personality, is unquestionably related to conscious limitation of use.[15]

The romantic idea of the noble savage living in a state of perfect equilibrium with nature is undoubtedly erroneous, as various studies suggest.[16] Nevertheless, the hunter's philosophy generally discourages him from eating himself out of house and home, as long as conditions remain fairly stable. This is not, unfortunately, the case with farming. The farmer has no choice but to set himself up in opposition to nature. Land must be cleared for cultivation, and weeds and pests, which would otherwise restore his fields to their original condition, must be vigorously suppressed. Domestic livestock must be controlled and confined, using force if necessary, to prevent them wandering off and reverting to a wild state, or being eaten by predators. The entire system, in fact, depends on the subjugation of nature, and the domination and manipulation of living creatures. It is likely that this new relationship with animals and nature, with its conflicting combination of intimacy and enslavement, generated intense feelings of guilt; guilt which was reinforced by an uncompromising environment that could ruin a crop overnight or decimate whole populations of livestock. Faced with this conflict, new ideologies were required; ideologies that absolved farming people from blame and enabled them to continue their remorseless programme of expansion and subjugation with a clear conscience. Different societies adopted different ideologies, some more expedient than others. But, if the record of civilization is anything to go by, it was the most ruthless cultures – the ones with the most effective distancing devices – who prospered most of all.

The ability to dominate and subdue animals and nature was so much a measure of success among the early civilizations that it often became a way of advertising personal, cultural or national prestige, like the building of pyramids, cathedrals or Olympic stadia. In many cases, this led to outstandingly gratuitous displays of brutality and despoliation. Throughout the ancient world, hunting, once an ordinary subsistence activity conducted in a restrained and respectful manner, became increasingly an opportunity for the ruling élite to publicize its dominance over lesser beings. Hunting became the peacetime equivalent of warfare and conquest; an excuse to indulge in conspicuous demonstrations of martial prowess, as well as providing a training ground for the acquisition of warrior attitudes and skills.[17] The kings or pharaohs of ancient Egypt, Babylonia, Assyria and Persia had wild animals placed in special walled enclosures or hunting preserves where they hunted them down at their leisure with the aid of dogs and chariots. They then had their noble exploits depicted in frescoes or in bas-relief on the walls of their palaces and tombs, so that everyone could be impressed by their obvious power.[18] Similarly, the ancient Greek soldier and historian, Xenophon, strongly advocated hunting as a method of cultivating manly virtues and military skills among young men.[19] The Greeks also developed a taste for ostentatious public displays involving animals. During the third century BC, a procession at Alexandria – the then cultural centre of the Hellenistic Empire – consisted of a file of people and animals that took an entire day to pass through the city's stadium. The spectacle included elephants, ostriches and wild asses harnessed to chariots, 2,400 dogs belonging to a variety of exotic breeds, 150 men bearing trees to which were attached birds and various arboreal mammals, a polar bear, 24 lions, 14 leopards, 16 cheetahs, 4 lynxes, a giraffe, a rhinoceros, and innumerable other wild and domestic birds and beasts.[20]

The Romans, of course, were notorious for their barbarous use of animals for entertainment. During their rise to imperial power, the citizens of Rome took an extraordinary delight in the sight of countless wild animals slaughtered by professional fighters (*bestiarii*), or goaded into fighting one another in the Circus Maximus and other arenas.* Like the Assyrians, they also staged theatrical hunts in which trained hunters (*venatores*) attacked and slew creatures before assembled crowds. Bears and bulls were chained together to

---

* The nearest modern equivalent of these Roman orgies of animal abuse are the bullfights of Spain and other Spanish-speaking countries. According to the only detailed anthropological study of bullfighting, it continues to celebrate masculine attributes – bravery, assertiveness, *machismo* – as well as expressing man's ability to dominate and subdue untamed nature (Marvin, 1988: 161–2).

fight; elephants, rhinoceroses, hippopotamuses, lions and leopards were fed intoxicants in order to excite them into a frenzy, and those that survived the ensuing carnage were shot from ringside seats by archers who paid specially for the privilege. The number of animals involved was prodigious. At 26 'beast hunts' organized by Caesar Augustus, 3,500 creatures were killed. In 2 BC, 260 lions were slaughtered at the Circus Maximus and 36 crocodiles at the Circus Flaminius. Gaius (Caligula) arranged a show at which 400 bears and 400 African beasts were killed and, not to be outdone, Nero allowed his personal bodyguard to murder 400 bears and 300 lions with javelins. The Emperor Trajan, however, held the record by ordering the public butchery of 11,000 beasts to celebrate his military victories in Dacia. Many Emperors enjoyed personal participation in the slaughter. Domitian apparently shot 100 assorted beasts with arrows at his Alban estate, and Commodus publicly despatched 100 bears, 6 hippopotamuses, 3 elephants, rhinoceroses, a tiger and a giraffe. According to one account, he also amused the multitude by shooting ostriches with special crescent-shaped arrowheads that were designed to decapitate the birds, while their headless bodies continued to run around.[21] Very occasionally these displays of rabid cruelty exceeded the otherwise limitless bounds of Roman decency. A staged hunt or *venatio* put on by Pompey the Great featured a performance in which eighteen African elephants were slaughtered by heavily armed gladiators. To Pompey's dismay, the elephants refused to defend themselves and the crowd actually took pity on the unfortunate beasts.[22]

Both the Greeks and the Romans regarded wild nature as a fearsome opponent to be mastered or avoided. The celebrations of nature found in classical literature are all of the cultivated, pastoral variety. In the first century BC, the poet Lucretius considered it a serious defect that so much of the world was 'greedily possessed by mountains and the forests of wild beasts'. Like most of his contemporaries, he derived considerable satisfaction from the belief that mankind had escaped from the horrors and the insecurity of pre-civilized life through the invention of agriculture, clothing, ships, city walls, roads, arms, laws and all the other trappings of modern civilization. The folklore of the period cultivated an image of the natural world as something mysterious, monstrous and frightening; a haven for menacing supernatural beings, man-eating ogres, werewolves and ghostly Wild Huntsmen. Even the forest god, Pan, had many sinister attributes, and was believed to engender intense fear or 'panic' in people who travelled in remote and desolate places.[23]

In medieval and early modern Europe, public participation in acts of gratuitous cruelty to animals was also commonplace. From the Middle Ages until at least the end of the eighteenth century – once again coinciding with the period of greatest agricultural and, later, imperial expansion – Europeans of all social classes revelled in the spectacle of bull-baiting, bear-baiting, cockfighting, dog-fighting and any number of exotic variations on the same theme. Meat animals were tortured to death in various ways, ostensibly to make their flesh more tender, while harmless and useful animals such as cats were mutilated, impaled on spits, plunged into boiling water, tossed onto bonfires, and hurled alive from the tops of tall buildings, and all in an atmosphere of extreme festive merriment.[24] On a visit to Kenilworth in 1575, Queen Elizabeth I of England was entertained by the sight of a pack of mastiffs baiting and being torn to pieces by a group of thirteen bears. When she was too old for horse-riding, her idea of fun was to sit in an ornately decorated bower and take pot-shots at confined herds of deer with a crossbow.[25] Meanwhile, hunting continued its triumphal progress to become the quintessential aristocratic pastime. King Louis XV of France seems to have spent most of his time on horseback chasing animals, and is credited with killing an incredible 10,000 red deer during his lifetime.[26]

With the sole exception of royal hunting preserves, uncultivated land was regarded as anathema, and throughout this period tremendous efforts were devoted to clearing forests, draining marshes, fens and waterlogged pastures and extending cultivation into heaths and mountainous areas.

According to Keith Thomas, 'agricultural improvement and exploitation were not just economically desirable; they were moral imperatives'. Wilderness was a symbol of chaos and the Devil, and its conversion to arable constituted a sort of pseudo-religious crusade designed to restore order and beauty to a world of nature that had turned against humanity at the Fall.[27]

Colonials and settlers carried similar values with them to the furthest outposts of European expansion. In colonial India and Africa, the flower of British manhood indulged in veritable orgies of big game slaughter. Roualeyn Gordon Cumming, perhaps the most celebrated of African white hunters, flouted all the conventional rules of sportsmanship. He had a habit of firing indiscriminately into herds of animals, shooting females and young as often as adult males, and he apparently thought nothing of wounding or laming animals and then leaving them to die. His personal accounts are full of callously dispassionate descriptions of animal death and suffering. After shooting a bull elephant no less than thirty-five times, he observed coolly how:

blood flowed from his trunk and all his wounds, leaving the ground behind him a mass of gore; his frame shuddered violently, his mouth opened and shut, his lips quivered, his eyes were filled with tears; he halted beside a thorny tree, and having turned right about he rocked forwards and backwards for a few seconds, and, falling heavily over, his ancient spirit fled.[28]

In the New World, such attitudes were honed and refined by the rigours of frontier life. The self-righteous Puritan divine, Cotton Mather, and other New Englanders, preached against wilderness as if it were an insult to God, and recommended its wholesale destruction as proof of religious conviction. In 1756, John Adams rejoiced in the fact that the American continent, once a 'dismal wilderness, the haunt of wolves, bears and more savage men' was now 'covered with fields of corn, orchards bending with fruit and the magnificent habitations of rational and civilized people'.[29] As the historian Roderick Nash observed:

frontiersmen acutely sensed that they battled wild country not only for personal survival but in the name of nation, race, and God. Civilizing the New World meant enlightening darkness, ordering chaos, and changing evil into good. In the morality play of westward expansion, wilderness was the villain, and the pioneer, as hero, relished its destruction. The transformation of wilderness into civilization was the reward for his sacrifices, the definition of his achievement, and the source of his pride.[30]

Anything that stood in the way of the inexorable march of civilization was hounded out of existence. The wolf, in particular, became the symbol of untamed nature and was persecuted and exterminated with implacable determination. Wolf bounty laws were passed in most areas and, within the space of about a century, the species was virtually eradicated in all but the states adjoining Canada. In the state of Montana alone, 80,730 wolves were bountied for $342,764 between the years of 1883 and 1918. The North American bison or buffalo fared even worse. Recognizing the buffalo's importance as prey to both the Indians and the wolves that inhabited the Great Plains, the authorities openly sanctioned their slaughter, and between 1850 and 1880 an incredible 75 million of these animals were massacred by hunters. Barry Lopez remarks that it is 'hard to look back on this period in American history and understand what motivated men to do what they did, to kill so thoroughly, so far in excess of what was necessary'.[31] Yet, in a sense, he answers his own question by arguing that the slaughter was an expression of Manifest Destiny, a symbol of the settler's God-given right to dominate and control his environment.

One of the best known exponents of this attitude was Theodore Roosevelt, a gung-ho wilderness hunter for whom the slaughter of big game represented an opportunity to flex and revitalize himself in manly, gladiatorial combat. To test his mettle to the full, Roosevelt focused most of his hunting efforts on large carnivores whose ferocity he unashamedly exaggerated. According to his own accounts, Roosevelt's bears were all claws, teeth and savagery, while his wolves and mountain lions were evil, slinking, criminal psychopaths.

Despite his cuddly association with teddy bears, Roosevelt displayed more than a streak of brutality toward animals in his writings. His journal for the summer of 1884 describes how he: 'broke the backs of two blacktail bucks with a single bullet', and how he shot a female grizzly bear and her cub, 'the ball going clean through him from end to end'.[32] Years later when he went on a 'collecting' safari in East Africa with his son Kermit, he shocked some of his British hosts by being 'utterly reckless in the expenditure of ammunition', and by killing far more than his quota of animals 'particularly the white rhinoceros, of which he and Kermit killed nine'. In all, the Roosevelt party shot 512 head of big game during the trip.[33]

Vestiges of this kind of pioneer brutality are still very much in evidence in many of the southern and western states of America. At country fairs live raccoons are baited to death with dogs for popular amusement, and dog-fighting, although illegal, is still widespread. One of the more bizarre events of this type is the 'Sweetwater Rattlesnake Round-Up', an annual get-together in Texas which draws up to 35,000 spectators, including many famous celebrities. The round-up, which may result in the capture and slaughter of up to 18,000 snakes in a single weekend, began in 1958 as a form of local pest control. Since then it has evolved into a commercial extravaganza featuring snake dancers and charmers, and demonstrations of snake catching, milking (for venom), butchering, skinning, cooking and devouring. A recent environmental assessment of the Sweetwater round-up describes it as: 'enmeshed in folklore and the cowboy mythos'. A contemporary rite of passage in which rattlesnakes are 'part of the uncivilized wilderness to be overcome and conquered. Destroying the killer rattler has become a way of prolonging the conquest of the frontier; continuing the excitement and adventure of the cowboy saga.'[34]

Prolonging 'the cowboy saga' is presumably also a factor promoting the survival of rodeos in the United States. In the popular imagination, the rodeo conjures up romantic images of the good old days of the Wild West, although nowadays it appears to be little more than an exercise in commercialized animal abuse. It is perhaps exaggerated to claim, as one author has, that the rodeo is 'the modern equivalent of the public hanging'. Nevertheless, these performances hinge on the violent subjugation of living animals, some of which are deliberately incited to frenzied violence by raking them with spurs, constricting the genital region with leather straps, or by thrusting an electric prod into the rectal area. At the same time they are often given bogus, malevolent names in order to deflect sympathy from their plight. Occasionally, they are maimed or killed, and many are forced to undergo the same terrifying ordeal several times a day.* Yet the rodeo is presented to the American public as harmless, red-blooded entertainment in which the cowboy – the epitome of wholesome, manly virtue – uses his courage and skill to overcome and subdue untamable, outlaw stock.[35] Doubtless the Romans employed similar fantasies to justify their activities in the Circus Maximus.

The obsession with ostentatious dominance and supremacy that characterizes post-Neolithic civilizations was not confined to their treatment of wilderness and other life forms. It also tended to engender an unsympathetic, oppressive and tyrannical attitude toward members of the human species. The hunter–gatherer's egalitarian view of the natural world, is generally reflected in the social organization of his group. Most of the people in the group enjoy roughly equal status, and there are no clearly definable leaders or bosses. Armed hostilities between neighbouring bands (unknown in some hunter–gatherer societies) are normally infrequent, spontaneous and short-lived affairs in which much of the fighting is ritualized or symbolic rather than deadly.[36] All this changed with the advent of agriculture.

It is common convention to look upon civilization in terms

---

* According to their own public relations information, the Professional Rodeo Cowboys Association has made an effort to improve the welfare of rodeo animals. Unfortunately, some of their claims do not inspire much confidence – e.g. 'most [rodeo animals] actually enjoy what they're doing'. The PRCA also admits that it oversees or sanctions only about a third of the rodeos which occur annually in the USA (PRCA, 1994: 3–4).

of its more benign attributes; its ingenuity and sophistication, its energy and creativity, and its tangible cultural achievements in the form of art, architecture and science.[37] It is easy to forget that, nine times out of ten, these refinements rested on foundations of extreme social inequality and organized military aggression. All of the world's great civilizations were based on strictly maintained social hierarchies dominated by some form of all-powerful ruling élite. The majority of the population lived as peasants or labourers who were compelled and, if necessary, physically coerced into doing what those in power wanted them to do. According to Harris, this kind of authoritarianism was an inevitable consequence of the need to intensify food production.[38] It was also a natural extension of the same techniques of self-deception that legitimized the enslavement and subordination of animals.

The great civilizations were also overtly war-like, and devoted vast resources to the business of invading, colonizing, subjugating and often enslaving less powerful cultures.[39] The expansionist nature of farming – the periodic pressure to exploit new land and new resources – gave birth to organized warfare and imperialism by bringing agricultural and pastoral peoples into constant competition with their neighbours.[40] And once again, the societies that emerged victorious from this struggle were the most ruthless, the most adept at dehumanizing their enemies and treating them as vermin or as beasts of burden. Throughout history, antipathy for wilderness, gratuitous cruelty to animals, and brutality toward people have walked hand in hand – all of them symptomatic of the battle for supremacy in an increasingly overcrowded, competitive and unforgiving world.*

The Romans, for example, notorious for their cruelty to wild beasts, dealt with their human opponents as if they were morally indistinguishable from animals. By the second century BC, Rome's military influence around the Mediterranean was

---

* Readers wishing to verify this assertion are recommended to read Yi-Fu Tuan's excellent review in *Dominance and Affection* (1984), and Jim Mason's remarkable synthesis, *An Unnatural Order* (1993).

so great that its citizens were able to live entirely off the proceeds of war, and the taxes paid by vassal states. It achieved this position through systematic acts of terror and brutality.[41] When the city of Corinth refused to cooperate with its Roman overlords in 146 BC, it was completely obliterated. When the Phoenician city of Carthage rebelled in the same year, it was immediately sacked, razed to the ground and its entire surviving population sold into slavery. The site of the city was then ploughed up and sown with salt so that no crops could be grown there again. Jews, Christians and other rebellious groups were taken back alive to Rome and publicly crucified, tortured, dipped in pitch and burned as 'human torches' or thrown, unarmed into the arenas to be mauled to death by animals in front of delighted crowds. The stronger male slaves spent the rest of their short lives chained to the oars of Roman galleys, or were incarcerated in special schools where they were trained to kill each other in gladiatorial combat.[42]

In Britain and Europe during the medieval and early modern period human life was similarly cheap. Public executions and torture were regular events that provided the masses with entertainment; and warfare, either at home or abroad, was virtually continuous. The working classes, the poor, the homeless, the mad and, to some extent, even women and children, had few if any rights, and were often regarded and treated as if they were less than human. In 1693, for example, Sir Thomas Pope Blount referred to the uneducated rabble as 'brutes' and argued that it was only 'by favour of a metaphor we call them men, for at best they are but Descartes's automata, moving frames and figures of men, and have nothing but their outsides to justify their titles to rationality'. Of course, depersonifying people in this way, and viewing them as beasts, conveniently allowed them to be treated accordingly. As Keith Thomas points out:

The ethic of human domination removed animals from the sphere of human concern. But it also legitimized the ill-treatment of those humans who were in a supposedly animal condition.[43]

The idea that certain human beings were inherently beast-like undoubtedly helped to justify Britain's participation in the Atlantic slave trade. Although few Englishmen had any first-hand experience of Africans, the majority unquestioningly assumed that these people were culturally and intellectually inferior, and some even regarded them as a sort of missing link between apes and men. Extraordinary myths circulated about their cannibal tendencies, their barbarism, their insatiable sexual appetites and their habit of copulating with apes. Needless to say, most of these stories were propagated by whites who had a vested interest in the trade.[44] Significantly, the British abolition of the slave trade in 1807 closely coincided with the earliest Parliamentary debates on cruelty to animals.

Treatment of Native Americans followed similar lines, although initial attitudes were more ambivalent. Many viewed them from the outset as bestial, subhuman cannibals, while others saw them as 'noble savages' inhabiting a 'Golden World'. Pietro's 1511 account of the second voyage of Columbus expounds the latter view. 'Mine and thine, the seeds of all mischief, have no place with them' he said:

they seem to live in the Golden World, without toil, living in open gardens, not entrenched with dykes, divided with hedges, or defended with walls. They deal truly with one another, without laws, without books, without judges. They take him for an evil and mischievous man which taketh pleasure in doing hurt to others.[45]

Following the settlement of Virginia and New England, a fragile truce existed between the colonists and their Indian neighbours. But repeated betrayals, homicides and violations of treaties, perpetrated primarily by the whites, produced a deepening atmosphere of mutual distrust. Eventually, alarmed by the growing incursion of settlers into their territory, the Indians retaliated by massacring a township of 347 people on the banks of the James River in 1622. This attack produced an immediate polarization of attitudes to the Indian which persisted for the next 300 years and guaranteed their virtual annihilation. Maniacal preachers, such as the Reverend Samuel Purchas, insisted that the Indians were:

bad people, having little of humanity but shape; ignorant of civility, of arts, of religion; more brutish than the beasts they hunt, more wild and unmanly than that unmanned wild country they range rather than inhabit; captivated also to Satan's tyranny, in foolish pieties, mad impieties, wicked idleness, busy and bloody wickedness.[46]

Soon many were recommending that the Indian territories be invaded, put to the sword and their inhabitants enslaved. Writing in 1625, Edward Waterhouse argued that military conquest was easier than trying to civilize the Indians by more subtle means. Victory, he said, could be easily effected:

by force, by surprise, by famine in burning their corn; by destroying and burning their boats, canoes and houses; by breaking their fishing weirs; by assailing them in their huntings, whereby they get the greatest part of their sustenance in winter; by pursuing and chasing them with our horses, and bloodhounds to draw after them and mastiffs to tear them.[47]

More than two centuries later, General Sheridan put the case more bluntly when he said 'the only good Indians I ever saw were dead'.

In Nazi Germany, Jews, Slavs, Gypsies, Poles, Bolsheviks, foreign workers and other 'undesirables' were relegated to the status of 'vermin' in order to justify their extermination. Hitler referred to the Jews as 'a pack of rats' and Himmler helped some of his soldiers cope with the horror of carrying out mass executions by telling them that 'bedbugs and rats have a life purpose . . . but this has never meant that man could not defend himself against vermin'. Nazi propaganda films, such as *Triumph of the Will* and *The Eternal Jew*, portrayed Jews and other supposedly 'degenerate people' as rats, or superimposed images of rats over pictures of Jews. Similarly, animal labels were used to legitimize biomedical research on humans, as if they were indistinguishable from laboratory animals.* At the

---

* Paradoxically, the Nazis also enacted some of the most comprehensive animal protection laws then in existence. This disturbing inversion of conventional moral priorities has been discussed in great detail elsewhere (Arluke & Sax, 1992; Arluke, 1993; Sax, 1993), and it provides a striking illustration of how, under certain circumstances, animals may be awarded a relatively higher moral status than some humans.

women's concentration camp at Ravensbrück, some of the Polish inmates, labelled 'rabbit girls', were given gas gangrene wounds or used as subjects for bone-grafting experiments. The concept of 'Untermenschen' or 'sub-human' was also used to classify non-Aryan peoples among the lowest forms of life. In one SS document, these 'sub-humans' are described as being 'mentally and emotionally on a far lower level than any animal' despite being physically identical to human beings.[48]

Nowadays, it is customary to regard the horrors of the Roman amphitheatre, or the callous brutality of eighteenth-century Englishmen, or the systematic extermination of Native Americans, buffaloes and wolves, or the more recent activities of the Nazis in Germany, the Khmer Rouge, the Bosnian Serbs or the Rwandan Hutus as the products of either individual or collective insanity. This is doubtless a healthy trend. Nevertheless, it should be remembered that, at the time, these pogroms and atrocities were readily justified by prevailing ideologies, and the men who instigated and led them were (or still are) often regarded as heroes or even demigods in their own lifetimes. The sad truth is that the ecological crisis that heralded the end of the last Ice Age, and the subsequent development of agriculture and animal domestication, propelled our ancestors into a highly competitive, violent and destructive period of history. A period that favoured individuals and societies who were able to dominate and subdue nature, living creatures and fellow humans without the traditions of moral restraint practised by their hunting forebears.

The reassuring idea that *civilization* represents *progress* is clearly in need of substantial and urgent revision. The historical success of so-called 'civilized' states arose not from their record of improving overall human well-being, but from their manifest ability to absorb or obliterate other less powerful groups.[49] In other words, since the Neolithic period, ruthlessness has been an adaptive characteristic, and the cultures that have come to dominate have done so by acting with remorseless detachment, or by manufacturing myths, fantasies, distortions and double standards to justify their high-handed, expansionist policies. The question is, will this particular brand of insanity always be adaptive; will it inevitably govern human affairs in the future? Or have we at last reached some sort of turning point?

Our species does appear to have driven itself into a corner. The distancing devices and the insane brutality that we have used so effectively to eliminate anyone or anything that stood in the way of progress have left us with a terrifying legacy. The few outposts of wilderness that remain are vanishing at an unprecedented rate, and most of the wild beasts that once threatened human livelihood are poised on the brink of extinction. In the so-called Developing World, over-population and over-intensive land use are reducing formerly fertile areas to desert or semi-desert at a rate of 140,000 square kilometres annually (an area larger than Greece or the state of Alabama), and roughly a billion people are close to starvation. The world's tropical forests, containing the highest diversity of living organisms, are being felled and burned at an even greater rate, and all but the most inaccessible will have vanished by the turn of the century.[50] To make matters worse, this process of destruction and decay is knowingly exacerbated by the intervention of the industrial superpowers. European, Asian and North American based multi-national companies are still encouraging these countries to produce cash crops and natural resources for export, instead of food for their own peoples.[51] Meanwhile, in the West, industrial waste-products are gradually polluting what is left of the natural environment, while over-enthusiastic use of agrochemicals is poisoning the ecosystem and the water supply.[52] The lingering obsession with intensifying agricultural output has also succeeded in generating either embarrassing and vastly expensive food surpluses, or bizarre, publicly funded 'set-aside' schemes* in

---

* According to a recent financial editorial in the London *Daily Telegraph*, farm subsidies – i.e. monies used to protect farmers from market forces, either by paying them more than the true market value for their produce, or for taking land out of production – cost European Community taxpayers £30 billion (or $47 billion US) in 1995. A figure which translates to a £1,300 surcharge on the annual food bill for the average family of four.

which farmers are actually paid *not* to produce anything.[53] Finally, the long-term global effects of greenhouse warming and atmospheric ozone depletion hang over us like an environmental Sword of Damocles waiting to drop.

Although military expansion can scarcely be justified on economic grounds any more, warfare, terrorism, racial and political violence, religious bigotry, social unrest and genocide appear to be increasing in proportion to economic hardship. The present century has probably been the most violent in history. The two world wars produced a horrific death toll and, although the great powers have not confronted each other for fifty years, wars and atrocities have been conducted almost continuously in other areas since 1945. And again, the West has exaggerated the problem by selling weapons at astronomical cost to poorer nations that can barely afford to feed themselves. The threat of global nuclear conflict may have diminished for the time being, but the inherent danger posed by the continued existence of such weapons of mass destruction is still enormous. The fictional deployment of nuclear or chemical weapons by fanatical terrorist groups has become a standard theme of political thrillers, and it seems only a question of time before life imitates art. Worst of all, perhaps, the old-fashioned, western, onwards-and-upwards view of infinite material progress has been exported to areas where such idealistic goals are not only physically unattainable, but potentially suicidal. As the historian John Roberts points out, 'if the West's ambiguous gifts to mankind include this, then along with our confidence and self-centredness, we shall have passed to the rest of the world a bias towards self-destruction'.[54] In short, we appear to be approaching the end of the line. We cannot expand; we seem unable to intensify production without wreaking further havoc, and the planet is fast becoming a wasteland.

Clearly, it matters a great deal how we set about tackling these problems. We can sit back and hope for a miracle; some unexpected revolution in science and technology that will revitalize the world's moribund economy and put us back on the road to prosperity. Such things are possible but most unlikely, and anyway we are not in a position to take risks. Or we can approach the crisis in our usual ruthless, bull-in-a-china-shop manner, and rely on some charismatic and, probably, psychopathic politician or general to lead us into a brave new world. This is precisely the approach that got us into our present mess, and it hardly seems appropriate for getting us out. Alternatively, we can take a lesson from our hunting and gathering predecessors, face up to our responsibilities, and attempt, as far as possible, to make reparation. Although, in many ways, this is the most difficult path to tread – and one which is likely to get more so as the crisis deepens and desperation mounts – there is at least some room for optimism. Over the past three centuries, ethical concern about the callous exploitation of animals, fellow humans and the natural world has been growing steadily, and within the past twenty to thirty years it has emerged as a force to be reckoned with.[55] Philanthropic societies and foundations, anti-nuclear campaigns, animal protection groups, international conservation organizations and national ecology parties have been gaining ground at an almost exponential rate, and all of them now exercise far greater social and political influence than ever before. Admittedly, their individual objectives differ, and they often appear to be in direct conflict with one another over a variety of issues.[56] Nevertheless, they are all, in a sense, a manifestation of the same thing: a collective sense of guilt and anger about what our species has done to itself, to other life forms and to the planet on which it lives, and a strong desire to make some form of restitution. Fortunately, as never before, we now have the knowledge and the expertise to do just that.

Cynics point out that this humanitarian and environmentalist movement is a largely western phenomenon, nurtured by material affluence and increasing urbanization. To some extent, this is undoubtedly true, but it makes no difference to the argument. As people have moved out of the rural environment and into towns and cities; as they have become less and less involved personally with the business of clearing wilderness, exterminating vermin and predators, and subduing and slaughtering livestock, they have inevitably begun to question the beliefs that legitimized this form of exploitation. Similarly, as Western economies have prospered, and the horrors of war have receded into the past, people have inevitably raised doubts about the value systems that justified social inequality and military adventurism. Just as inevitably, the edifice of myths and falsehoods on which these beliefs and values were based has started to crumble. Moral scruples of this kind failed to gain ground in the past because they were incompatible with the prevailing direction in which society was moving. Ruthlessness and self-deception prevailed because these were what circumstances demanded. Now they demand something different. Necessity always was and always will be the mother of invention, and necessity dictates that we alter our perceptions and reject the anthropocentric and ethnocentric man-versus-nature, us-against-them philosophies that have dominated the past 10,000 years of human history.

It would be overly idealistic to predict that humans will one day live on terms of absolute equality with other species or, indeed, with each other. It would also, in my view, be unrealistic to imagine that we can hope to achieve global vegetarianism, or a complete end to the economic utilization of animals or the natural environment. Paradise, in this sense, cannot be regained because it never really existed. Nevertheless, it is clear that we cannot go on treating the world and its contents like some gigantic supermarket. Economic, political or religious ideologies that promote unrestrained exploitation are dangerous. They threaten our survival not only by the irreparable damage they cause, but also by denying, suppressing or corrupting feelings and morality. Fortunately, and thanks largely to our past excesses, ethical arguments based on the principles of empathy and altruism, and economic objectives based on long-term human interests, are, at long last, beginning to converge. We can but hope that out of this union a sane and responsible compromise will emerge.

This book began with a paradox, a paradox exemplified

by a society in which a dispassionate, utilitarian attitude to factory-farmed livestock coexists with affectionate and sympathetic relationships with domestic pets. I have argued that popular beliefs about why people keep pets are often erroneous, and that they should be replaced with the notion that personifying animals and befriending them is a normal and natural human characteristic, and one that can be emotionally fulfilling. I have also suggested that negative and disparaging views of pet-keeping and other non-anthropocentric activities persist precisely because affectionate or empathic perceptions of animals or nature are incompatible with our unsympathetic treatment of economically useful species. The harsh and implacable philosophies that underwrite this treatment were spawned in an era in which moral reservations about ruthless exploitation were out of place. It was necessary, instead, to suppress empathic feeling, to cultivate detachment, to conceal the facts or distort them, and, where possible, to shift the blame for what was happening away from the perpetrators. Above all, it was necessary to fabricate an image of humanity – more especially western humanity –that was separate and apart from the rest of creation, sacred and superior, answerable to no one but God and, more recently, Mammon.

But like all illusions, this concept is fragile and easily refuted. The discoveries of Copernicus, Galileo and Newton were certainly damaging, and a mortal blow was struck by Darwin's theory of evolution by natural selection. Even pets, strange as it may seem, may have played their part in its downfall.* For when we elevate companion animals to the status of persons; when we empathize with them and acknowledge their resemblance to ourselves, it becomes obvious that the notion of human moral superiority is a phantom: a dangerous, egotistical myth that currently

---

* This is more than mere idle conjecture. Recent studies suggest that young adults with relatively sympathetic attitudes towards animals and the environment are also more likely to report having experienced positive relationships with pets during childhood (Paul & Serpell, 1993; Serpell & Paul, 1994).

threatens our survival. Ironically, as the forerunner of animal domestication, pet-keeping led us into our present, destructive phase of history. Perhaps, by making us more aware of our biological affinities with animals and the natural world, it will help to lead us out again.

Notes

PREFACE TO THE CANTO EDITION

1 Sheppard, 1986, p. 49.
2 Gould, 1987, p. 20; Ritvo, 1986.
3 Klein, 1995, p. 19.
4 *Ibid.*, p. 23.

I OF PIGS AND PETS

 1 Murdock, 1968, p. 13.
 2 Lee & DeVore, 1968, pp. 3–12.
 3 Harris, 1978, pp. 9–12.
 4 Leakey & Lewin, 1977, pp. 123, 144–5.
 5 Lee & DeVore, 1968, pp. 9–12.
 6 Clutton-Brock, 1981, pp. 34ff; Davis, 1982, pp. 697–700; Malek, 1993, pp. 45–54.
 7 Zeuner, 1963, pp. 436–54; Harris, 1969, pp. 3–15.
 8 Murdock, 1968, p. 13.
 9 Burkert, 1983, p. 20; Campbell, 1984, p. 54; Ingold, 1994, pp. 9–12; Levine, 1971, pp. 426–7.
10 Signoret *et al.*, 1975, pp. 295–329.
11 Pond, 1983, pp. 78–87.
12 Carnell, 1983, p. 36.
13 Hammond *et al.*, 1983, pp. 139–68.
14 S. Baxter, 1984, pp. 478–80.
15 Pond, 1983, pp. 78–87.
16 S. Baxter, 1984, pp. 478–80.
17 *Ibid.*
18 Singer, 1984, pp. 120–3.
19 *Ibid.*
20 *Ibid.*, p. 158.
21 Schleifer, 1985, pp. 63–4.

22 North, 1983, p. 63.
23 Hammond *et al.*, 1983, pp. 63–4.
24 The procedure described is based on the author's own observations in a modern intensive pig-breeding unit.
25 Hammond *et al.*, 1983, pp. 63–4.
26 English *et al.*, 1977, p. 15.
27 Mason, 1985, pp. 89–107.
28 FEDIAF, 1994; Messent & Horsfield, 1985, pp. 9–17.
29 Rowan, 1995, p. 111.
30 Serpell, 1983a, pp. 80–4.
31 Mugford, 1981, pp. 295–315; Mugford, 1995, pp. 140–52.
32 AVMA, 1992; PFI, 1994.
33 PFMA, 1995.
34 Soulsby & Serpell, 1988, pp. 15–24.
35 D. N. Baxter, 1984a, pp. 29–36.
36 Lockwood, 1995, p. 132.
37 Gershman *et al.*, 1994, p. 913.
38 Studman, 1983, p. 8.
39 *Ibid.*
40 Galbraith & Barrett, 1986, pp. 634–44; Kirkwood, 1987, pp. 98–9; Baxter & Leck, 1984, pp. 185–97.
41 Baxter & Leck, 1984, *ibid.*
42 Soulsby & Serpell, 1988, pp. 46–8; Wolfensohn, 1981, pp. 404–7.
43 Soulsby & Serpell, 1988, pp. 42–5; Hubrecht, 1995, pp. 180–2.
44 Arkow, 1994, pp. 202–5.
45 Passmore, 1983.
46 Serpell, 1995, pp. 248–50.
47 Pond, 1983, pp. 78–87.

2 SUBSTITUTES FOR PEOPLE

1 Ritvo, 1987, pp. 85–6; Ritvo, 1988, pp. 16–21; Kete, 1994, pp. 1–3.
2 Pritchard, 1988, pp. 98–101.
3 Humphrey, 1983, p. 98.
4 Halliday, 1922, pp. 151–4.
5 Midgley, 1983, p. 116.
6 Katcher, 1981, p. 46.
7 Tuan, 1984, pp. 2–6.
8 Menninger, 1951, p. 57.
9 Szasz, 1968, p. 91; Klein, 1995, p. 23.
10 Herzog & Galvin, 1992, pp. 80–2.

11 Anon., 1984b.
12 Chorlton, 1984; Anon., 1984a.
13 Smith, 1984.
14 Voith, 1981, pp. 271–94.
15 Cormier, 1990; Schroer, 1994.
16 Windeler, 1983.
17 Phillips, 1982; Schroer, 1994.
18 Windeler, 1983.
19 Plumb, 1975, pp. 46–7.
20 Tannenbaum, 1995, pp. 314–15.
21 Windeler, 1983.
22 Penny, 1976, p. 302.
23 Jesse, 1866, vol. II, p. 26.
24 Keddie, 1977, pp. 21–5.
25 Anon., 1985b; Fogle, 1983, pp. 99–123.
26 Ryncarson, 1978, pp. 550–5.
27 Worth & Beck, 1981, pp. 282–98; Clifton, 1993, p. 8.
28 Szasz, 1968, p. 90; Arluke & Sax, 1992, pp. 15–19.
29 Dale-Green, 1966, pp. 169–72.
30 Carson, 1972, p. 10
31 Ellis, 1928, pp. 76–88; Menninger, 1951, pp. 64–71.
32 Thomas, 1983, p. 39.
33 Kinsey *et al.*, 1948, pp. 669–78; Kinsey *et al.*, 1953, pp. 505–9.
34 Ellis, 1928, pp. 76–88.
35 Beck & Katcher, 1983, p. 75.
36 Russel, 1956, pp. 19–35.
37 Cameron *et al.*, 1966, pp. 884–6; Cameron & Mattson, 1972, p. 286.
38 Thomas, 1983, p. 119; Serpell & Paul, 1994, p. 135.
39 Hume, 1957, p. 33; ten Bensel, 1984, pp. 9–12; Linzey, 1987, pp. 14–16.
40 Rothschild, 1985.
41 Simon, 1984, pp. 226–40.
42 Brown *et al.*, 1972, pp. 957–8.
43 Lee, 1976.
44 Kidd & Feldmann, 1981, pp. 867–75.
45 Martinez & Kidd, 1980, p. 381.
46 Friedmann *et al.*, 1984, pp. 300–8.
47 Griffiths & Bremner, 1977, pp. 1333–40; Franti *et al.*, 1980, pp. 143–9; Messent & Horsfield, 1985, pp. 9–17.
48 Messent & Horsfield, *ibid.*
49 Serpell, 1981, pp. 651–4.
50 Kidd *et al.*, 1983, pp. 719–29.

3 INSTRUMENTS OF FOLLIE

1 Speke, 1863, pp. 288–92.
2 Dixie, 1931; Burkhardt, 1960, p. 84.
3 Ash, 1927, p. 59.
4 Burkhardt, 1960, p. 84; Osgood, 1975, pp. 945–7.
5 O'Callaghan, 1983; Anon., 1983.
6 O'Callaghan, 1983.
7 Watts, 1985.
8 Halliday, 1922, pp. 151–4.
9 *Ibid.*
10 Penny, 1976, p. 298.
11 Merlen, 1971, pp. 63–4.
12 Thomas, 1983, pp. 40, 100–20.
13 Labarge, 1980, p. 184.
14 Armstrong, 1973, p. 7.
15 Harwood, 1928, p. 40.
16 Ritchie, 1981, p. 64.
17 *Ibid.*
18 Labarge, 1980, p. 184.
19 Szasz, 1968, p. 17.
20 Jesse, 1866, vol. II, p. 228.
21 Thomas, 1983, p. 40.
22 Szasz, 1968, p. 17.
23 Ritchie, 1981, p. 118.
24 Thomas, 1983, pp. 100–20; Ritvo, 1988, pp. 16–21.
25 Pepys, 1970–76, vol. I, p. 54, vol. IV, p. 99, vol. IX, p. 308.
26 Savishinsky, 1983, pp. 116–18.
27 Tuan, 1984.
28 *Ibid.*
29 Jesse, 1866, vol. I, p. 115.
30 Ritchie, 1981, p. 119.
31 Norman, 1985.
32 Jesse, 1866, vol. I, p. 334.
33 Ouida, 1891, p. 317.
34 Midgley, 1983, p. 116.
35 D. N. Baxter, 1984b, p. 200.
36 Woosey, 1983.
37 Pond, 1983, pp. 78–87; Soulsby & Serpell, 1988, p. 61.
38 Thomas, 1971, pp. 521–4.
39 Ewen, 1933, pp. 71–4; Thomas, 1971, pp. 530–1; Rowland, 1990, p. 173.
40 Rosen, 1969, pp. 30–2; Hole, 1977, pp. 38–53; Ewen, 1933.

41 Rosen, 1969, pp. 109–10.
42 Hole, 1977, p. 40.
43 Natoli, 1984.
44 Orwell, 1947, p. 10.

4 PETS IN TRIBAL SOCIETIES

1 Hernandez, 1651, pp. 295–6.
2 Elmendorf & Kroeber, 1960, p. 114.
3 Beck, 1983, p. 238.
4 Galton, 1883, pp. 246–7.
5 Fernandez, 1937, pp. 57–8.
6 Quoted in Gosse, 1881, p. 331.
7 Juan & Ulloa, 1760, p. 426.
8 Galton, 1883, pp. 246–7.
9 Roth, 1970, pp. 551–6.
10 Fleming, 1984, pp. 145–6.
11 Wilbert, 1972, pp. 96–7.
12 Basso, 1973, p. 21.
13 Hugh-Jones, 1985, personal communication.
14 Galton, 1883, pp. 251–3.
15 Lumholtz, 1884, pp. 178–9.
16 Cipriani, 1966, pp. 80–1.
17 Evans, 1937, p. 64.
18 Galton, 1883, pp. 251–3.
19 Luomala, 1960, pp. 202–3.
20 Jesse, 1866, vol. 1, pp. 298–9.
21 Luomala, 1960, pp. 202–3
22 Serpell, 1988a, p. 15.
23 Lévi-Strauss, 1966, pp. 203–16; Tambiah, 1969, pp. 452–3.
24 Leach, 1964, pp. 23–63
25 Harris, 1978, pp. 120–1.
26 Sahlins, 1976, pp. 174–5.
27 Laughlin, 1968, p. 309.
28 Downs, 1964, p. 23.
29 Lumholtz, 1884, pp. 178–9.
30 Cipriani, 1966, pp. 80–1.
31 Harrison, 1965, pp. 67–86.
32 Meggit, 1965, pp. 17–18.
33 Hugh-Jones, 1985, personal communication.
34 Singer, 1968, pp. 270–9.
35 Linton, 1936, pp. 428–9.
36 Erikson, 1987, pp. 105–40.

37 Galton, 1883, pp. 243–71; Reed, 1954, pp. 1629–39; Serpell, 1989a, pp. 17–19.
38 Davis & Valla, 1978, pp. 608–10.

5 A CUCKOO IN THE NEST

1 Wilson, 1975, pp. 375–7.
2 *Ibid.*
3 Voith, 1984, pp. 147–56.
4 Lorenz, 1943, pp. 235–409; Csermely & Mainardi, 1984, pp. 1–19; Sternglanz *et al.*, 1977, pp. 108–15; Berman, 1980, pp. 668–95.
5 Gardner & Wallach, 1965, pp. 135–42.
6 Jolly, 1972, pp. 228–9.
7 Lorenz, 1943, pp. 235–409.
8 Gould, 1979, p. 35.
9 Wyllie, 1981.
10 Berryman *et al.*, 1985, pp. 659–61.
11 Beck & Katcher, 1983, p. 60.
12 Hirsh-Pasek & Treiman, 1982, pp. 229–37.
13 Katcher, 1983, pp. 519–31.
14 Voith, 1984, pp. 154–5.
15 Salmon & Salmon, 1983, pp. 29–36.
16 Smith, 1983, pp. 29–36; Hart, 1995, p. 165.
17 Fogle, 1983, p. 148
18 Simon, 1984, pp. 226–40.
19 Quoted in Katcher, 1983, p. 524.
20 Jesse, 1866, vol. II, p. 228.
21 Orwell, 1947, p. 10.
22 Lawrence, 1989, p. 14.
23 Juan & Ulloa, 1760, p. 426.
24 Lumholtz, 1889, pp. 178–9; Galton, 1883, pp. 251–2.
25 Basso, 1973, p. 21.
26 Savishinsky, 1983, pp. 119–20.
27 Hose & McDougall, 1912, pp. 70–1.
28 Briggs, 1970, p. 70; Savishinsky, 1983, p. 119.
29 Messent & Serpell, 1981, pp. 17–20; Coppinger & Schneider, 1995, pp. 38–9.
30 Price, 1984, p. 22.
31 Frank & Frank, 1982, pp. 507–22.
32 Dawkins & Krebs, 1979, pp. 489–511.
33 Mugford, 1983, pp. 40–4.
34 Wolfensohn, 1981, pp. 404–7.
35 Wilson, 1975, pp. 364–8.

6 PETS AS PANACEA

1 Levinson, 1964, p. 243.
2 Levinson, 1978, pp. 1031–8; Levinson, 1969, pp. 158–78; Levinson, 1980, pp. 63–81; Levinson, 1962, pp. 61–5; Levinson, 1972.
3 Dale-Green, 1966, p. 133.
4 *Ibid.*, pp. 135–6; Toynbee, 1973, p. 123.
5 Schmitt, 1983, pp. 2–6, 145–55.
6 Toynbee, 1973, p. 123.
7 Jesse, 1866, vol. ii, p. 229.
8 McCulloch, 1983, pp. 410–26.
9 Bucke, 1903, p. 510.
10 Sherick, 1981, p. 193.
11 Bossard, 1944, pp. 408–13.
12 Corson & O'Leary Corson, 1980, pp. 83–110.
13 Hines, 1983, pp. 7–17.
14 Lee, 1983, pp. 24–5.
15 Salmon & Salmon, 1982; Hogarth Scott *et al.*, 1983, pp. 4–6.
16 Winkler *et al.*, 1989, pp. 216–23.
17 Katcher, 1982, pp. 2–8; Beck & Katcher, 1984, pp. 410–27.
18 Serpell, 1983c.
19 Bustad, 1980; Bustad & Hines, 1981, pp. 787–810; McCulloch, 1983, pp. 410–26; Ross, 1983, pp. 26–39.
20 Friedmann *et al.*, 1980, pp. 307–12.
21 *Ibid.*
22 Friedmann *et al.*, 1984, pp. 300–8.
23 Katcher, 1981, pp. 41–67.
24 Sebkova, 1977.
25 Friedmann *et al.*, 1983, pp. 461–5.
26 Katcher *et al.*, 1984, pp. 171–8.
27 Robb & Stegman, 1983, pp. 277–82; Lago *et al.*, 1989, pp. 25–34; Watson & Weinstein, 1993, pp. 135–8.
28 Ory & Goldberg, 1983, pp. 303–17.
29 Stallones *et al.*, 1990.
30 Akiyama *et al.*, 1987, pp. 187–93; Bolin, 1987, pp. 26–35.
31 Siegel, 1990, pp. 1081–6.
32 Anderson *et al.*, 1992, pp. 298–307.
33 Serpell, 1990b, pp. 1–7; Serpell, 1991, pp. 717–20.
34 Katcher *et al.*, 1983, pp. 351–9.
35 McCaul & Malott, 1984, pp. 516–33.
36 Lockwood, 1983, pp. 64–71.
37 *Ibid.*

38 Katcher, 1983, p. 253; Beck & Katcher, 1983, p. 60.
39 Klein, 1995, pp. 20–1.
40 McNiell Taylor, 1983, pp. 71–2.
41 Messent, 1983, pp. 37–46.
42 Eddy *et al.*, 1988, pp. 39–45; Mader *et al.*, 1989, pp. 1529–34.
43 Gubser, 1965, pp. 292–3.
44 ten Bensel, 1984, pp. 2–14; Hume, 1957, p. 29; Midgley, 1983, p. 51; Rothschild, 1985; Serpell & Paul, 1994, pp. 136–7.
45 Kellert & Felthous, 1985, pp. 1113–29.
46 Katcher, 1982, pp. 2–8.
47 Prioleau *et al.*, 1983, pp. 275–310.
48 Mugford, 1980, pp. 111–12; Serpell, 1989b, pp. 111–12.

7 HEALTH AND FRIENDSHIP

1 Serpell, 1989b, pp. 112–18.
2 Wilson, 1975, pp. 565–75; Bertram, 1978, pp. 64–96.
3 Schachter, 1959, pp. 1–2.
4 Perlman & Peplau, 1981, pp. 31–56.
5 Schachter, 1959, pp. 9–10.
6 Larson *et al.*, 1982, pp. 40–53.
7 Schachter, 1959, pp. 6–8.
8 Bowlby, 1980, p. 9.
9 Perlman & Peplau, 1981, pp. 31–56.
10 Panksepp *et al.*, 1980, pp. 473–87.
11 Keverne, 1988, pp. 127–41; Keverne & Kendrick, 1994, pp. 47–56; Martel *et al.*, 1993, pp. 307–21.
12 Seyfarth & Cheney, 1984, p. 541.
13 Keverne *et al.*, 1989, pp. 155–61.
14 Panksepp *et al.*, 1980, pp. 473–87.
15 Martin, 1977, pp. 3–35.
16 Bowlby, 1980, p. 9.
17 Cross, 1984, pp. 10–11.
18 Martin, 1977, pp. 3–35.
19 Perlman & Peplau, 1981, pp. 31–56.
20 Duck, 1983, p. 7; House *et al.*, 1988, pp. 540–5.
21 Eriksen, 1994, pp. 201–9.
22 Syme, 1974, pp. 1043–5; Bloom *et al.*, 1978, pp. 867–94; Lynch, 1977, p. 45.
23 Duck, 1983, p. 15.
24 Herbert & Cohen, 1993, pp. 472–86; Vilhjalmson, 1993, pp. 331–42; Eriksen, 1994, pp. 207–9; Ader *et al.*, 1995, pp. 99–103.

25 Martin, in press.
26 Selye, 1957.
27 Ader *et al.*, 1995, pp. 99–103.
28 Levine, 1983, pp. 15–26; Jemmot & Locke, 1984, pp. 78–108; Sklar & Anisman, 1981, pp. 369–406.
29 Anisman & Zarcharko, 1982, pp. 89–157.
30 Schachter, 1959, pp. 26, 133.
31 Lynch, 1977, p. 101.
32 *Ibid.*, p. 123.
33 *Ibid.*, p. 102.
34 Eriksen, 1994, pp. 201–9.
35 Kielcolt-Glaser *et al.*, 1993, pp. 395–409; Malarkey *et al.*, 1993, pp. 41–51.
36 Bloom *et al.*, 1978, pp. 867–94.
37 Bowlby, 1980, p.9; Ader *et al.*, 1995, pp. 99–103; Martin, in press.
38 Laudenslager *et al.*, 1983, p. 568.
39 Cohen *et al.*, 1992, pp. 301–4.
40 Nerem *et al.*, 1980, pp. 1475–6.
41 Gross & Siegel, 1982, pp. 2010–12.
42 Lynch & McCarthy, 1966, p. 81; Lynch, 1977, pp. 167–80.
43 Duck, 1983, pp. 17–31.
44 Mehrabian, 1971, pp. 64–73; Argyle, 1975, pp. 211–50; Serpell, 1989b, pp. 115–16.
45 Perlman & Peplau, 1981, pp. 31–56.
46 Katcher, 1981, p. 50.

8 FOUR-LEGGED FRIENDS

1 Messent & Serpell, 1981, pp. 5–22.
2 Mech, 1970, pp. 133–6; Fox, 1975a, pp. 410–36.
3 Messent & Serpell, 1981, p. 16.
4 *Ibid.*
5 Borchelt, 1983b, pp. 45–61; Voith, 1981, pp. 271–94.
6 Kleimann, 1967, pp. 279–88.
7 Hinde, 1979, pp. 55–138; Serpell, 1989b, pp. 112–13.
8 Lynch, 1977, pp. 217–30.
9 Serpell, 1983b, pp. 57–63.
10 Borchelt, 1983a, pp. 187–96; Serpell & Jagoe, 1995, pp. 89–90.
11 Mathes & Deuger, 1982, pp. 251–354.
12 Anon., 1985a.
13 Schenkel, 1967, pp. 319–30.

14 Moelk, 1979, pp. 164–224; Karsh & Turner, 1988, pp. 163–76; Deag *et al.*, 1988, pp. 26–8.
15 Katcher, 1981, pp. 41–67.
16 Pitcairn & Eibl-Eibesfeldt, 1976, pp. 81–113; Kendon & Ferber, 1973, pp. 591–668.
17 Mech, 1970, pp. 80–95.
18 Moelk, 1979, pp. 164–224.
19 Ginsberg, 1976, pp. 59–79.
20 Bullowa, 1979, pp. 1–62.
21 Mitchell, 1972, pp. 53–67.
22 Kendon, 1967, pp. 22–63; Argyle, 1975, pp. 211–50.
23 Papousek & Papousek, 1977, pp. 63–85.
24 Fraiberg, 1979, pp. 149–69.
25 Argyle, 1975, pp. 211–50.
26 London, 1903, p. 165.
27 Mech, 1970, pp. 80–95.
28 Kendon, 1967, pp. 591–668; Argyle, 1975, pp. 211–50.
29 Bolwig, 1962, pp. 167–92.
30 Quoted in D. N. Baxter, 1984a, pp. 29–36.
31 Katcher, 1983, p. 523.
32 Thorpe, 1972, pp. 27–47.
33 Serpell, 1989b.
34 Weiss, 1972, pp. 17–26.

9 THE MYTH OF HUMAN SUPREMACY

1 Messent & Horsfield, 1985, pp. 7–17.
2 Serpell, 1981, pp. 651–4.
3 Levinson, 1975, pp. 8–9; Fox, 1975b, p. 38.
4 Damon, 1977, p. 140.
5 Messent & Horsfield, 1985.
6 Beck & Katcher, 1983, p. 25; Hart, 1995, p. 165.
7 Thomas, 1983, pp. 17–22.
8 Hume, 1957, pp. 8–29; Shell, 1988, pp. 137–40.
9 Hume, 1957, pp. 8–29; Farrington, 1969, pp. 75–81; Sorabji, 1993.
10 Sorabji, 1993.
11 Augustine, *City of God*, 1.20.
12 Sorabji, 1993, pp. 170–207.
13 *Ibid.*, p. 3.
14 Aquinas, *Summa contra Gentiles*, II, 112.
15 Aquinas, *Summa Theologiae*, q.102, a. 6.
16 Hume, 1957, pp. 8–29.
17 Lovell, 1979, pp. 1–4.

18 Thomas, 1983, pp. 17–22.
19 Maehle, 1994, pp. 86–7.
20 Singer, 1984, pp. 217–23.
21 Cohn, 1975, pp. 22–31.
22 Lovell, 1979, pp. 1–4.
23 Singer, 1984, pp. 217–23.
24 Hume, 1957, pp. 8–29.
25 Thomas, 1983, pp. 17–22.
26 Schmitt, 1983, pp. 2–6.
27 Hamilton, 1981, pp. 21–2; Cohn, 1975, pp. 54–9; Spencer, 1993, pp. 136–71.
28 Sprenger & Kramer, 1968; Hughes, 1952, p. 101; Rosen, 1969, pp. 30–2; Cohn, 1975, pp. 99–102.
29 Hyde, 1916, p. 711; Evans, 1906, pp. 146–57.
30 Thomas, 1983, p. 39.
31 Sprenger & Kramer, 1968.
32 Ritchie, 1981, pp. 69–70.
33 Jesse, 1866, vol. 1, p. 228; Thomas, 1983, p. 40.
34 Rowland, 1972, p. 161.
35 Thomas, 1983, pp. 38–41.
36 *Ibid.*, pp. 105–9.
37 Rosen, 1969, p. 32.
38 Lovell, 1979, pp. 1–4.
39 Singer, 1984, pp. 217–23.
40 Thomas, 1983, pp. 143–5.
41 Singer, 1984, pp. 8–9.
42 Singer, 1984, pp. 217–23.
43 Thomas, 1983, pp. 143–5.
44 Midgley, 1983, p. 51.
45 Carson, 1972, p. 24.
46 Maehle, 1994, p. 99; Ritvo, 1988, pp. 16–21; Serpell & Paul, 1994, pp. 134–6.
47 Jenkins, 1978, pp. 24–48.
48 Burrow, 1968, pp. 11–48; Oldroyd, 1980, pp. 20–3; Thomas, 1983, p. 130.
49 Oldroyd, 1980, pp. 193–203.
50 *Ibid.*
51 Irvine, 1955, p. 130.
52 Oldroyd, 1980, p. 247.
53 *Ibid.*
54 Lockwood, 1989, pp. 41–56; Kennedy, 1992, pp. 1–7.
55 Griffin, 1976, p. 69.
56 Hume, 1962, pp. 104–5.

10 KILLER WITH A CONSCIENCE

1 Singer, 1984, p. 219.
2 Thomas, 1983, p. 41.
3 *Ibid.*, pp. 25–30.
4 Thomas, 1971, pp. 3–24.
5 Anon., 1984c; Hediger, 1964, pp. 166–9.
6 Hediger, 1965, pp. 291–300; Serpell, 1989b, pp. 118–24.
7 Hughes, 1973, pp. 10–11; McFarland, 1981, pp. 16–17; Kennedy, 1992, pp. 1–8.
8 McFarland & McFarland, 1969, p. 14.
9 Zahn-Waxler *et al.*, 1985, pp. 21–39.
10 Sharefkin & Ruchlis, 1974, pp. 37–44.
11 Blanchard, 1982, pp. 586–91; Tucker, 1989, pp. 167–72; Serpell & Paul, 1994, pp. 137–8; Soulsby & Serpell, 1988, p. 26.
12 Ingold, 1994, p. 18.
13 Lockwood, 1989, pp. 41–56; Midgley, 1983, pp. 141–2.
14 Humphrey, 1983, pp. 5–9.
15 Cheney & Seyfarth, 1990, p. 303.
16 Garcia, 1981, p. 151.
17 Humphrey, 1983, p. 6.
18 Burghardt & Herzog, 1989, pp. 139–48; Kellert, 1989, pp. 20–3; Driscoll, 1992, pp. 32–8; Serpell & Paul, 1994, p. 128.
19 Serpell, 1985, pp. 112–16; Serpell, 1995, p. 253.
20 Menninger, 1951, pp. 58–9.
21 Hultzkrantz, 1965, p. 303.
22 Levine, 1971, pp. 420–7.
23 Serpell & Paul, 1994, p. 132.
24 Levine, 1971, pp. 420–7.
25 Durkheim, 1971, pp. 336–40; Lévi-Strauss, 1966.
26 Benedict, 1929, pp. 19–43; Hallowell, 1926, pp. 3–12.
27 Campbell, 1984, pp. 81–122.
28 Levine, 1971, p. 432.
29 Tuan, 1984, p. 91.
30 Turnbull, 1976, pp. 90–6; Campbell, 1984, pp. 105–6.
31 Henriksen, 1973, pp. 26–39.
32 Lopez, 1978, pp. 90–3; Speck, 1935, pp. 79–92.
33 Ingold, 1980, p. 282.
34 Fortes, 1967, pp. 1–15; Frazer, 1913, pp. 410–16; Morris, 1995, p. 311.
35 Hugh-Jones, 1985, personal communication.
36 Menninger, 1951, p. 52.
37 Furer-Haimendorf, 1943, p. 231.
38 Burkert, 1983, pp. 13–14; Hultzkrantz, 1965, p. 304.

39 Serpell & Paul, 1994, p. 131.
40 Tuan, 1984, p. 95; Cartmill, 1993, pp. 232–7.
41 MacKenzie, 1988, pp. 45–51.
42 Anderson, 1985, pp. 19–22.
43 Levine, 1971, p. 432.
44 Campbell, 1984, pp. 53–6.
45 *Ibid.*
46 Hallowell, 1926, pp. 31–163.
47 Campbell, 1984, pp. 147–54.

II LICENSED TO KILL

1 Midgley, 1983, p. 113.
2 Maccoby, 1982, p. 50.
3 Lorenz, 1954, p. vii.
4 Rothschild, 1985.
5 Vialles, 1994, pp. 61–5.
6 Evans-Pritchard, 1940, pp. 16–50.
7 Ingold, 1980, pp. 97–101.
8 *Ibid.*, pp. 282–5.
9 Carnell, 1983, pp. 97–8; Mason, 1985, pp. 89–107.
10 Carnell, 1981, p. 25.
11 Jesse, 1866, vol. 1, p. 299.
12 Serpell, 1995, pp. 248–50.
13 Luomala, 1960, pp. 202–4.
14 Papashvily & Papashvily, 1954, p. 126.
15 Briggs, 1970, pp. 289–94; Henriksen, 1973, p. 19; Savishinsky, 1983,
   p. 119.
16 Arluke, 1988, pp. 105–9; Serpell, 1989c, p. 165.
17 Herzog, 1988, pp. 473–4.
18 Thomas, 1983, p. 294; Schleifer, 1985, p. 65.
19 Sahlins, 1976, pp. 175–6.
20 Ezard, 1984.
21 Walker, 1984.
22 Dennis, 1981.
23 Russel, 1956, pp. 29–35.
24 Lopez, 1978, p. 145.
25 *Ibid.*, p. 139.
26 Mills, 1984, pp. 28–31.
27 Crisler, 1958; Mowatt, 1963; Mech, 1970.
28 Hyde, 1916, pp. 696–730.
29 Luomala, 1960, p. 218.
30 Serpell, 1995, pp. 250–2.

31 Tuan, 1984, p. 95; Serpell, 1989c, p. 162.
32 Maccoby, 1982, p. 8.
33 Fisher, 1983, pp. 132–7.
34 Thomas, 1983, pp. 294–5.
35 Schleifer, 1985, pp. 68–9.
36 Girard, 1977, pp. 6–7; Gill, 1982, pp. 85–9.
37 Harris, 1978, pp. 99–125.
38 Burkert, 1983, pp. 3–12; Sorabji, 1993, p. 180.
39 Sorabji, 1993, p. 171.
40 Burkert, 1983, pp. 3–12.
41 Sorabji, 1993, pp. 170–94.
42 Cartmill, 1993, p. 38.
43 Sorabji, 1993, p. 180.
44 Burkert, 1983, pp. 3–12.
45 Leach, 1964; Sahlins, 1976, pp. 170–9.

12 THE FALL FROM GRACE

1 Harris, 1978, pp. 9–14; Cohen & Armalagos, 1984, pp. 585–601; Cohen, 1989, pp. 116–22.
2 Lee, 1969, pp. 47–9.
3 Harris, 1978, pp. 9–14.
4 Allen, 1980, pp. 11–16.
5 UNEP, 1989, p. 345.
6 Cohen, 1989, p. 132.
7 *Ibid.*, pp. 140–1.
8 Harris, 1978, pp. 9–14; Martin, 1967, pp. 75–120.
9 Cohen, 1977, pp. 1–17.
10 *Ibid.*
11 Bayliss-Smith, 1982, pp. 13–14.
12 Roberts, 1985, p. 265.
13 Wenzel, 1991, pp. 134–41; Ingold, 1994, pp. 12–15.
14 Nelson, 1986, p. 211.
15 *Ibid.*, pp. 220–1.
16 Cohen, 1977; Martin, 1967, pp. 75–120; Martin, 1978.
17 MacKenzie, 1988, pp. 8–13.
18 Burkert, 1983, pp. 42–3.
19 Cartmill, 1993, p. 31.
20 Tuan, 1984, p. 74.
21 Toynbee, 1973, pp. 21–3; Goodenough, 1979, pp. 108–10.
22 Cartmill, 1993, p. 41.
23 Nash, 1982, pp. 9–12.
24 Serpell, 1988b, p. 156; Cartmill, 1993, p. 104.

25 Thomas, 1983, pp. 143–9.
26 Cartmill, 1993, p. 66.
27 Thomas, 1983, pp. 254–8.
28 MacKenzie, 1988, pp. 98–9.
29 Lopez, 1978, p. 142.
30 Nash, 1982, p. 24.
31 Lopez, 1978, p. 183.
32 Mighetto, 1991, pp. 36–7.
33 MacKenzie, 1988, p. 162.
34 Weir, 1992, p. 123.
35 Carson, 1972, pp. 156–82.
36 Harris, 1978, pp. 34–5; Lesser, 1968, pp. 92–6.
37 Roberts, 1985, p. 9.
38 Harris, 1978, pp. 69–82.
39 McNeill, 1983, pp. 1–28.
40 Harrison, 1973, pp. 26–7.
41 Tuan, 1984, p. 11.
42 Goodenough, 1979, pp. 15, 103–8.
43 Thomas, 1983, pp. 41–50.
44 Barker, 1978, pp. 15–58.
45 Arber, 1885, p. 78.
46 Purchas, 1906, p. 231.
47 Kingsbury, 1933, pp. 541–57.
48 Arluke & Sax, 1992, pp. 24–6.
49 Cohen, 1989, pp. 141–2.
50 Allen, 1980, pp. 11–16; UNEP, 1989, p. 345; WCMC, 1994; Dregne
    et al., 1991.
51 Roberts, 1985, p. 400.
52 Allen, 1980, pp. 11–16; UNEP, 1989, pp. 39–190.
53 Brown, 1985; Body, 1984; pp. 1–15.
54 Roberts, 1985, pp. 10, 427.
55 Thomas, 1983, pp. 300–303; White, 1967, pp. 1203–7.
56 Serpell, 1992, pp. 27–39.

Bibliography

Ader, R. L., Cohen, N. & Felten, D. (1995). Psychoneuroimmunology: interactions between the nervous system and the immune system. *The Lancet*, 345, 99–103.

Akiyama, H., Holtzman, J. M. & Britz, W. E. (1987). Pet-ownership and health status during bereavement. *Omega: Journal of Death and Dying*, 17, 187–93.

Allen, R. (1980). *How to Save the World*. Gland: IUCN.

Anderson, J. K. (1985). *Hunting in the Ancient World*. Berkeley, CA: University of California Press.

Anderson, W. P., Reid, C. M. & Jennings, G. L. (1992). Pet ownership and risk factors for cardiovascular disease. *Medical Journal of Australia*, 157, 298–301.

Anisman, H. & Zarcharko, R. M. (1982). Depression: the predisposing influence of stress. *The Behavioural and Brain Sciences*, 5, 89–157.

Anon. (1983). Dogs slaughtered. *Guardian*, 30 November.

(1984a). Catty tug-of-love over Marmaduke costs £6,000. *Guardian*, 15 June.

(1984b). Will they let Lucy live? *Daily Mail*, 23–6 January.

(1984c). Gorilla wept for pet cat. *Daily Telegraph*, 10 January.

(1985a). Inquest warning on pets and babies. *The Times*, 24 April.

(1985b). Man's suicide after dog died. *Guardian*, 13 February.

Arber, E. (1885). *The First Three English Books on America*. Birmingham: Turnbull & Spears, Edinburgh.

Argyle, M. (1975). *Bodily Communication*. London: Methuen.

Arkow, P. (1994). A new look at pet 'over-population'. *Anthrozoös*, 7, 202–5.

Arluke, A. (1988). Sacrificial symbolism in animal experimentation: object or pet. *Anthrozoös*, 2, 98–117.

(1993). Author's response: good to hate with – animal and Nazi symbols then and now. *Anthrozoös*, 6, 98–107.

(1994). Managing emotions in an animal shelter. In *Animals & Human Society: Changing Perspectives,* eds. A. Manning & J. A. Serpell, pp. 145–65. London: Routledge.

Arluke, A. & Sax, B. (1992). Understanding Nazi animal protection and the Holocaust. *Anthrozoös,* 5, 6–31.

Armstrong, E. A. (1973). *Saint Francis: Nature Mystic.* Berkeley, CA: University of California Press.

Ash, E. C. (1927). *Dogs: Their History and Development.* London: Ernest Benn.

AVMA (1992). *The Veterinary Service Market for Companion Animals.* Schaumberg, IL: American Veterinary Medical Association Center for Information Management.

Barker, A. J. (1978). *The African Link: British Attitudes to the Negro in the 17th and 18th Centuries.* London: Frank Cass.

Basso, C. B. (1973). *The Kalapalo Indians of Central Brazil.* New York: Holt, Rinehart & Winston.

Baxter, D. N. (1984a). The deleterious effects of dogs on human health: dog-associated injuries. *Community Medicine,* 6, 29–36.

(1984b). The deleterious effects of dogs on human health: 3. Miscellaneous problems and a control programme. *Community Medicine,* 6, 198–203.

Baxter, D. N. & Leck, I. (1984). The deleterious effects of dogs on human health: 2. Canine zoonoses. *Community Medicine,* 6, 185–97.

Baxter, S. (1984). *Intensive Pig Production: Environmental Management and Design.* London: Collins.

Bayliss-Smith, T. D. (1982). *The Ecology of Agricultural Systems.* Cambridge University Press.

Beck, A. M. (1983). Animals in the city. In *New Perspectives on Our Lives with Companion Animals,* eds. A. H. Katcher & A. M. Beck, pp. 237–43. Philadelphia: University of Pennsylvania Press.

(1984). Population aspects of animal mortality. In *Pet Loss and Human Bereavement,* pp. 42–8. Ames: Iowa State University Press.

Beck, A. M. & Katcher, A. H. (1983). *Between Pets and People.* New York: G. P. Putnam.

(1984). A new look at pet-facilitated therapy. *Journal of the American Veterinary Medical Association,* 184, 410–27.

Benedict, R. F. (1929). The concept of the guardian spirit in North America. *Memoirs of the American Anthropological Association,* 29, 3–93.

Berman, P. W. (1980). Are women more responsive than men to the young? A review of developmental and situational variables. *Psychological Bulletin,* 88, 668–95.

Berryman, J. C., Howells, K. & Lloyd-Evans, M. (1985). Pet owner attitudes to pets and people. *Veterinary Record*, 117, 659–61.

Bertram, B. C. R. (1978). Living in Groups. In *Behavioural Ecology: An Evolutionary Approach*, eds. N. B. Davies & J. R. Krebs, pp. 64–96. Oxford: Blackwell.

Blanchard, J. S. (1982). Anthropomorphism in beginning readers. *The Reading Teacher*, 35 (5), 586–91.

Bloom, B. L., Asher, S. J. & White, S. W. (1978). Marital disruption as a stressor: a review and analysis. *Psychological Bulletin*, 85, 867–94.

Body, R. (1984) *Agriculture: The Triumph and the Shame*. London: Temple Smith.

Bolin, S. E. (1987). The effects of companion animals during conjugal bereavement, *Anthrozoös*, 1, 26–35.

Bolwig, N. (1962). Facial expression in primates with remarks on parallel development in certain carnivores. *Behaviour*, 22, 167–92.

Borchelt, P. L. (1983a). Separation-elicited behaviour problems in dogs. In *New Perspectives on Our Lives with Companion Animals*, eds. A. H. Katcher & A. M. Beck, pp. 187–96. Philadelphia: University of Pennsylvania Press.

   (1983b). Aggressive behaviour of dogs kept as companion animals: classification and influence of sex, reproductive status and breed. *Applied Animal Ethology*, 10, 45–61.

Bossard, J. H. S. (1944). The mental hygiene of owning a dog. *Mental Hygiene*, 28, 408–13.

Bowlby, J. (1980). *Loss, Sadness and Depression: Attachment and Loss*, vol. III. London: Hogarth Press.

Briggs, J. (1970). *Never in Anger: Portrait of an Eskimo Family*. Cambridge, MA: Harvard University Press.

Brown, D. (1985). Fatted calves may milk mountain. *Guardian*, 28 September.

Brown, L. T., Shaw, T. G. & Kirkland, K. D. (1972). Affection for people as a function of affection for dogs. *Psychological Reports*, 31, 957–8.

Bucke, W. F. (1903). Cyno-psychoses: children's thoughts, reactions and feelings toward pet dogs. *Pedagological Seminary (Journal of Genetical Psychology)*, 10, 459–513.

Bullowa, M. (1979). Pre-linguistic communication: a field for scientific research. In *Before Speech*, ed. M. Bullowa, pp. 1–62. Cambridge University Press.

Burkert, W. (1983). *Homo necans*, trans. P. Bing. Berkeley: University of California Press.

Burkhardt, G. & Herzog, H. (1989). Animals, evolution, and ethics. In *Perceptions of Animals in American Culture*, ed. R. J. Hoage, pp. 129–51. Washington, DC: Smithsonian Institution Press.

Burkhardt, V. R. (1960). *Chinese Creeds and Customs*, vol. III. Hong Kong: South China Morning Post Ltd.

Burrow, J. W. (1968). Editor's introduction. In Darwin, C., *The Origin of Species by Means of Natural Selection*, pp. 11–48. London: Penguin Books.

Bustad, L. K. (1980). *Animals, Aging and the Aged*. Minneapolis: University of Minnesota Press.

Bustad, L. K. & Hines, L. M. (1981). The human–companion animal bond and the veterinarian. *Veterinary Clinics of North America: Small Animal Practice*, 11, 787–810.

Cameron, P. & Mattson, M. (1972). Psychological correlates of pet-ownership. *Psychological Reports*, 30, 286.

Cameron, P., Conrad, C., Kirkpatrick, D. & Bateen, R. (1966). Pet ownership and sex as determinations of stated affect towards others and estimates of other's regard of self. *Psychological Reports*, 19, 884–6.

Campbell, J. (1984). *The Way of the Animal Powers*. London: Times Books.

Carnell, P. (1981). An economic appraisal of less intensive systems in egg production and breeding pigs. In *Alternatives to Intensive Husbandry Systems*, pp. 21–31. Potter's Bar: Universities Federation for Animal Welfare.

(1983). *Alternatives to Factory Farming*. London: Earth Resources Research Ltd.

Carson, G. (1972). *Men, Beasts and Gods: A History of Cruelty and Kindness to Animals*. New York: Charles Scribner's Sons.

Cartmill, M. (1993). *A View to a Death in the Morning: Hunting and Nature through History*. Cambridge, MA: Harvard University Press.

Cheney, D. L. & Seyfarth, R. M. (1990). *How Monkeys See the World: Inside the Mind of Another Species*. Chicago University Press.

Chorlton, P. (1984). A marmalade cat among the legal pigeons. *Guardian*, 5 March.

Cipriani, L. (1966). *The Andaman Islanders*. London: Weidenfeld & Nicolson.

Clifton, M. (1993). Demographics of the shelter dog population. *Animal People*, November, pp. 7–8.

Clutton-Brock, J. (1981). *Domesticated Animals from Early Times*. London: Heinemann, British Museum of Natural History.

(1995). Origins of the dog: domestication and early history. In *The Domestic Dog: Its Evolution, Behaviour & Interactions with People*, ed. J. A. Serpell, pp. 7–20. Cambridge University Press.

Cohen, M. N. (1977). *The Food Crisis in Prehistory*. New Haven, CT: Yale University Press.

(1989). *Health and the Rise of Civilization*. New Haven, CT: Yale University Press.

Cohen, M. N. & Armelagos, G. J. (1984). Palaeopathology at the origins of agriculture: editors' summation. In *Palaeopathology at the Origins of Agriculture*, eds. M. N. Cohen & G. J. Armelagos, pp. 585–601. New York: Academic Press.

Cohen, S. & Williamson, G. M. (1991). Stress and infectious disease. *Psychological Bulletin*, 109, 5–24.

Cohen, S., Kaplan, J. R., Cunnick, J. E., Manuck, S. B. & Rabin, B. S. (1992). Chronic social stress, affiliation, and cellular immune response in nonhuman primates. *Psychological Science*, 3, 301–4.

Cohn, N. (1975). *Europe's Inner Demons: An Enquiry Inspired by the Great Witch-hunt*. New York: Basic Books.

Coppinger, R. & Schneider, R. (1995). The evolution of working dogs. In *The Domestic Dog: Its Evolution, Behaviour & Interactions with People*, ed. J. A. Serpell, pp. 21–47. Cambridge University Press.

Cormier, W. R. (1990). Fashion mavens swaddle pets in matching designer outfits. *Los Angeles Times*, 4 June.

Corson, S. A. & O'Leary Corson, E. (1980). Pet animals as nonverbal communication mediators in psychotherapy in institutional settings. In *Ethology and Nonverbal Communication in Mental Health*, eds. S. A. Corson & E. O'Leary Corson, pp. 83–110. Oxford: Pergamon Press.

Crisler, L. (1958). *Arctic Wild*. New York: Harper Brothers.

Cross, M. (1984). A case for legal heroin. *New Scientist*, 1422, 10–11.

Csermely, D. & Mainardi, D. (1984). Infant signals. In *The Behaviour of Human Infants*, eds. A. Oliverio & M. Zapella, pp. 1–19. New York: Plenum Press.

Dale-Green, P. (1966). *Dog*. London: Rupert Hart-Davis.

Damon, W. (1977). *The Social World of the Child*. San Francisco: Jossey-Bass Publishers.

Davis, S. (1982). The taming of the few. *New Scientist*, 1322, 697–700.

Davis, S. J. M. & Valla, F. R. (1978). Evidence for the domestication of the dog 12,000 years ago in the Natufian of Israel. *Nature*, 276, 608–10.

Dawkins, R. & Krebs, J. R. (1979). Arms races between and within species. *Proceedings of the Royal Society of London*, ser. B, 205, 489–511.

Deag, J. M., Manning, A. & Lawrence, C. (1988). Factors influencing the mother–kitten relationship. In *The Domestic Cat: The Biology of Its Behaviour*, eds. D. C. Turner & P. P. G. Bateson, pp. 23–40. Cambridge University Press.

Dennis, N. (1981). Those animal crackers. *The Sunday Telegraph*, 2 August.

Dixie, A. C. (1931). *The Lion Dog of Peking*. London: Peter Davies.

Downs, J. F. (1964). *Animal Husbandry in Navaho Society and Culture*. Berkeley: University of California Press.

Dregne, H., Kassas, M. & Rosanov, B. (1991). A new assessment of the world status of desertification. *UNEP Desertification Control Bulletin*, 20.

Driscoll, J. W. (1992). Attitudes towards animal use. *Anthrozoös*, 5, 32–8.

Duck, S. (1983). *Friends for Life*. Brighton: Harvester Press.

Durkheim, E. (1971). *The Elementary Forms of the Religious Life*, 7th edn. London: George Allen & Unwin.

Eddy, J., Hart, L. A. & Boltz, R. P. (1988). The effects of service dogs on social acknowledgements of people in wheelchairs. *Journal of Psychology*, 122, 39–45.

Ellis, H. (1928). *Studies in the Psychology of Sex*, vol. v. Philadelphia: F. A. Davis.

Elmendorf, W. W. & Kroeber, K. L. (1960). The structure of Twana culture with comparative notes on the structure of Yurok culture. *Washington University Research Studies*, Monograph 2, 28.

English, P. R., Smith, W. J. & MacLean, A. (1977). *The Sow-Improving Her Efficiency*. Ipswich: Farming Press.

Eriksen, W. (1994). The role of social support in the pathogenesis of coronary heart disease: a literature review. *Family Practice*, 11, 201–9.

Erikson, P. (1987). De l'apprivoisement à l'approvisionnement: chasse, alliance et familiarisation en Amazonie Amérindienne. *Techniques et Cultures*, 9, 105–40.

Evans, E. P. (1906). *The Criminal Prosecution and Capital Punishment of Animals*. London: Heinemann.

Evans, I. H. N. (1937). *The Negritos of Malaysia*. Cambridge University Press.

Evans-Pritchard, E. E. (1940). *The Nuer*. Oxford University Press.

Ewen, C. l'Estrange. (1933). *Witchcraft and Demonianism*. London: Heath Cranton.

Ezard, J. (1984). Getting meat's image off the hook. *Guardian*, 30 November.

Farrington, B. (1969). *Aristotle: Founder of Scientific Philosophy*. New York: Praeger.

FEDIAF (European Pet Food Federation) (1994). Information sheet.

Fernandez, G. de Oviedo. (1937). Natural history of the West Indies. In *Studies in the Romance Languages and Literature*, vol. xxiii, trans. S. A. Stoudemire. Chapel Hill: University of North Carolina Press.

Fisher, M. P. (1983). Of pigs and dogs: Pets as produce in three societies. In *New Perspectives on Our Lives with Companion Animals*, eds. A. H. Katcher & A. M. Beck, pp. 132–7. Philadelphia: University of Pennsylvania Press.

Fleming, P. (1984). *Brazilian Adventure*. London: Penguin Books.

Fogle, B. (1983). *Pets and their People*. London: Collins.

Fortes, M. (1967). Totem and taboo. *Proceedings Royal Anthropological Institute of Great Britain and Northern Ireland for 1966*, 5–22.

Fox, M. W. (1975a). The behaviour of cats. In *The Behaviour of Domestic Animals*, ed. E. S. E. Hafez, pp. 410–36. London: Baillière Tindall.

(1975b). Pet–owner relations. In *Pet Animals and Society*, ed. R. S. Anderson, p. 37–53. London: Baillière Tindall.

Fraiberg, S. (1979). Blind infants and their mothers: an examination of the sign system. In *Before Speech*, ed. M. Bullowa, pp. 149–69. Cambridge University Press.

Frank, H. & Frank, M. G. (1982). On the effects of domestication on canine social development and behaviour. *Applied Animal Ethology*, 8, 507–22.

Franti, C. E., Kraus, J. F., Borhani, N. O., Johnson, S. L. & Tucker, S. D. (1980). *Journal of the American Veterinary Medical Association*, 176, 143–9.

Frazer, J. G. (1913). *Aftermath: A Supplement to the Golden Bough*. London: Macmillan.

Friedmann, E., Katcher, A. H., Lynch, J. J. & Thomas, S. A. (1980). Animal companions and one-year survival of patients after discharge from a coronary care unit. *Public Health Report*, 95, 307–12.

Friedmann, E., Katcher, A. H., Thomas, S. A., Lynch, J. J. & Messent, P. R. (1983). Interaction and blood pressure: influence of animal companions. *Journal of Nervous and Mental Disease*, 171, 461–5.

Friedmann, E., Katcher, A. H., Eaton, M. & Berger, B. (1984). Pet ownership and psychological status. In *The Pet Connection*, eds. R. K. Anderson, B. L. Hart & L. A. Hart, pp. 300–8. Minneapolis: Center to Study Human–Animal Relationships and Environments, University of Minnesota.

Furer-Haimendorf, C. (1943). *The Chenchus*. London: Macmillan.

Galbraith, N. S. & Barrett, N. J. (1986). Emerging zoonoses. *Journal of Small Animal Practice*, 27, 621–46.

Galton, F. (1883). *Inquiry into Human Faculty and its Development.* London: Macmillan.

Garcia, J. (1981). Tilting at the paper mills of academe. *American Psychologist,* 36, 151.

Gardner, B. T. & Wallach, L. (1965). Shapes of figures identified as a baby's head. *Perceptual and Motor Skills,* 20, 135–42.

Gershman, K. A., Sacks, J. J. & Wright, J. C. (1994). Which dogs bite? A case-control study of risk factors. *Pediatrics,* 93, 913–17.

Gill, S. D. (1982). *Beyond the Primitive: The Religions of Non-literate Peoples.* Englewood Cliffs: Prentice-Hall.

Ginsberg, B. E. (1976). Evolution of communication patterns in animals. In *Communicative Behaviour and Evolution,* eds. M. E. Hahn & E. C. Simmel, pp. 59–79. New York & London: Academic Press.

Girard, R. (1977). *Violence and the Sacred,* trans. P. Gregory. Baltimore: Johns Hopkins University Press.

Goodenough, S. (1979). *Citizens of Rome.* London: Hamlyn.

Gosse, P. H. (1881). *A Naturalist's Sojourn in Jamaica.* London: Longman, Brown, Green & Longman's.

Gould, S. J. (1979). Mickey Mouse meets Konrad Lorenz. *Natural History,* 88, 30–6.

(1987) Animals and us. *New York Review of Books,* 25 June, pp. 20–5.

Grandin, T. (1988). Behavior of slaughter plant and auction employees toward the animals. *Anthrozoös,* 1, 205–13.

(1994). Farm animal welfare during handling, transport, and slaughter. *Journal of the American Veterinary Medical Association,* 204, 372–7.

Griffin, D. (1976). *The Question of Animal Awareness.* New York: Rockefeller University Press.

Griffiths, A. D. & Bremner, A. (1977). Survey of dog and cat ownership in Champaign County, Illinois. *Journal of the American Veterinary Medical Association,* 170, 1333–40.

Gross, W. B. & Siegel, P. B. (1982). Socialization as a factor in resistance to infection, feeding efficiency, and response to antigen in chickens. *American Journal of Veterinary Research,* 43, 2010–12.

Gubser, N. J. (1965). *The Nunamiut Eskimos: Hunters of the Caribou.* New Haven, CT: Yale University Press.

Halliday, W. R. (1922). Animal pets in ancient Greece. *Discovery,* 3, 151–4.

Hallowell, A. I. (1926). Bear ceremonialism in the northern hemisphere. *American Anthropologist,* 28, 1–175.

Halverson, J. (1976). Animal categories and terms of abuse. *Man* (N.S.), 11, 505–16.

Hamilton, B. (1981). *The Medieval Inquisition*. London: Edward Arnold.

Hammond, J., Bowman, J. C. & Robinson, T. J. (1983). *Hammond's Farm Animals*, 5th edn. London: Edward Arnold.

Harris, D. R. (1969). Agricultural systems, ecosystems and the origins of agriculture. In *The Domestication and Exploitation of Plants and Animals*, eds. P. J. Ucko & G. W. Dimbleby, pp. 3–15. London: Duckworth.

Harris, M. (1978). *Cannibals and Kings*. London: Collins.

Harrison, R. (1973). *Warfare*. Minneapolis: Burgess Publishing Co.

Harrison, T. (1965). Three 'secret' communication systems among Borneo nomads (and their dogs). *Journal of the Malay Branch of the Royal Asiatic Society*, 38, 67–86.

Hart, L. A. (1995). Dogs as human companions: a review of the relationship. In *The Domestic Dog: Its Evolution, Behaviour & Interactions with People*, ed. J. A. Serpell, pp. 161–78. Cambridge University Press.

Harwood, D. (1928). *Love for Animals and How it Developed in Great Britain*. New York: Columbia University Press.

Hediger, H. (1964). *Wild Animals in Captivity*. New York: Dover Publications.

(1965). Man as a social partner of animals and vice-versa. *Symposium of the Zoological Society of London*, 14, 291–300.

Heiman, M. (1965). Psychoanalytic observations on the relationship of pet and man. *Veterinary Medicine/Small Animal Clinician*, 16 July, pp. 713–18.

Henriksen, G. (1973). *Hunters in Barrens: The Naskapi on the Edge of the White Man's World*. St. Johns: Institute of Social and Economic Research, Memorial University of Newfoundland.

Herbert, T. B. & Cohen, S. (1993). Depression and immunity: a meta-analytic review. *Psychological Bulletin*, 113, 472–86.

Hernandez, F. (1651). Historiae animalium et mineralium Novae Hispaniae. In *Rerum Medicarum Novae Hispaniae*, eds. N. A. Recchi & J. T. Lynceus, pp. 295–6. Rome, 4to.

Herzog, H. A. (1988). The moral status of mice. *American Psychologist*, 43, 473–4.

Herzog, H. A. & Galvin, S. (1992). Animals, archetypes and popular culture: tales from the tabloid press. *Anthrozoös*, 5, 77–92.

Hinde, R. A. (1979). *Towards Understanding Relationships*. New York: Academic Press.

Hines, L. M. (1983). Pets in prisons. *California Veterinarian*, 37, 7–17.

Hirsh-Pasek, K. & Treiman, R. (1982). Doggerel: Motherese in a new context. *Journal of Child Language*, 9, 229–37.

Hogarth Scott, S., Salmon, I. & Lavelle, R. (1983). A dog in residence. *People–Animals–Environment*, 1, 4–6.

Hole, C. (1977). *Witchcraft in England*. London: Batsford.

Hose, C. & McDougall, M. B. (1912). *The Pagan Tribes of Borneo*, vol. II. London: Macmillan.

House, J. S., Landis, K. R. & Umberson, D. (1988). Social relationships and health. *Science*, 241, 540–5.

Hubrecht, R. (1995). The welfare of dogs in human care. In *The Domestic Dog: Its Evolution, Behaviour & Interactions with People*, ed. J. A. Serpell, pp. 179–98. Cambridge University Press.

Hughes, A. (1973). Anthropomorphism, teleology, animism and personification – why they should be avoided. *Science and Children*, 10 (7), 10–11.

Hughes, P. (1952). *Witchcraft*. London: Longmans, Green & Co.

Hultzkrantz, A. (1965). Type of religion in the arctic hunting cultures. In *Hunting and Fishing*, ed. H. Hvarfner, p. 303. Lulea: Norrbottens Museum, Sweden.

Hume, C. W. (1957). *The Status of Animals in the Christian Religion*. Potter's Bar: Universities Federation for Animal Welfare.

    (1962). *Man and Beast*. Potter's Bar: Universities Federation for Animal Welfare.

Humphrey, N. (1983). *Consciousness Regained*. Oxford University Press.

Hyde, W. W. (1916). The prosecution and punishment of animals and lifeless things in the Middle Ages and modern times. *University of Pennsylvania Law Review*, 64, 696–730.

Ingold, T. (1980). *Hunters, Pastoralists and Ranchers*. Cambridge University Press.

    (1994). From trust to domination: an alternative history of human–animal relations. In *Animals & Human Society: Changing Perspectives*, eds. A. Manning & J. A. Serpell, pp. 1–22. London: Routledge.

Irvine, W. (1955). *Apes, Angels and Victorians*. Lanham, MD: University Press of America.

Jemmot, J. B. & Locke, S. E. (1984). Psychosocial factors, immunologic mediation, and human susceptibility to infectious diseases. *Psychological Bulletin*, 95, 78–108.

Jenkins, A. C. (1978). *The Naturalists*. London: Book Club Associates.

Jesse, G. R. (1866). *Researches into the History of the British Dog*, vols. I & II. London: Robert Hardwicke.

Jolly, A. (1972). *The Evolution of Primate Behaviour*. London: Macmillan.

Juan, G. & Ulloa, A. De. (1760). *Voyage to South America*, vol. I. London.

Karsh, E. & Turner, D. C. (1988). The human–cat relationship. In *The Domestic Cat: The Biology of Its Behaviour,* eds. D. C. Turner & P. P. G. Bateson, pp. 159–78. Cambridge University Press.

Katcher, A. H. (1981). Interactions between people and their pets: form and function. In *Interrelations between People and Pets,* ed. B. Fogle, pp. 41–67. Springfield, IL: Charles Thomas.

(1982). Are companion animals good for your health? *Aging,* 331, 2–8.

(1983). Man and the living environment: an excursion into cyclical time. In *New Perspectives on Our Lives with Companion Animals,* eds. A. H. Katcher & A. M. Beck, pp. 519–31. Philadelphia: University of Pennsylvania Press.

Katcher, A. H., Friedmann, E., Beck, A. M. & Lynch, J. J. (1983). Looking, talking and blood pressure: the physiological consequences of interaction with the living environment. In *New Perspectives on Our Lives with Companion Animals,* eds. A. H. Katcher & A. M. Beck, pp. 351–9. Philadelphia: University of Pennsylvania Press.

Katcher, A. H., Segal, D. D. S. & Beck, A. M. (1984). Contemplation of an aquarium for the reduction of anxiety. In *The Pet Connection,* eds. R. K. Anderson, B. L. Hart & L. A. Hart, pp. 171–8. Minneapolis: CENSHARE, University of Minnesota.

Keddie, K. M. D. (1977). Pathological mourning after the death of a domestic pet. *British Journal of Psychiatry,* 131, 21–5.

Kellert, S. R. (1989). Perceptions of animals in America. In *Perceptions of Animals in American Culture,* ed. R. J. Hoage, pp. 5–24. Washington, DC: Smithsonian Institution Press.

Kellert, S. R. & Felthous, A. R. (1985). Childhood cruelty toward animals among criminals and non-criminals. *Human Relations,* 38, 1113–29.

Kendon, A. (1967). Some functions of gaze direction in social interaction. *Acta Psychologica,* 26, 22–63.

Kendon, A. & Ferber, A. (1973). A description of some human greetings. In *Comparative Ecology and Behaviour of Primates,* eds. R. P. Michael & J. H. Crook, pp. 591–668. New York & London: Academic Press.

Kennedy, J. S. (1992). *The New Anthropomorphism.* Cambridge University Press.

Kete, K. (1994). *The Beast in the Boudoir: Petkeeping in Nineteenth-Century Paris.* Berkeley, CA: University of California Press.

Keverne, E. B. (1988). Central mechanisms underlying the neural and neuroendocrine determinants of maternal behaviour. *Psychoneuroendocrinology,* 13, 127–41.

Keverne, E. B. & Kendrick, K. M. (1994). Maternal behaviour in sheep and its neuroendocrine regulation. *Acta Paediatrica Supplement*, 397, 47–56.

Keverne, E. B., Martensz, E. B. & Tuite, B. (1989). Beta-endorphin concentrations in cerebrospinal fluid of monkeys are influenced by grooming relationships. *Psychoneuroendocrinology*, 14, 155–61.

Kidd, A. H. & Feldmann, B. M. (1981). Pet ownership and self-perceptions of older people. *Psychological Reports*, 48, 867–75.

Kidd, A. H., Kelly, H. T. & Kidd, R. M. (1983). Personality characteristics of horse, turtle, snake and bird owners. *Psychological Reports*, 52, 719–29.

Kidd, A. H. & Kidd, R. M. (1989). Factors in adults' attitudes towards pets. *Psychological Reports*, 65, 903–10.

Kiecolt-Glaser, J. K. & Glaser, R. (1991). Stress and immune function in humans. In *Psychoneuroimmunology*, 2nd edn, eds. R. Ader, D. L. Felten & N. Cohen, pp. 849–68. New York: Academic Press.

Kiecolt-Glaser, J. K., Malarkey, W. B., Chee, M., Newton, T., Cacioppo, J. T., Mao, H.-Y. & Glaser, R. (1993). Negative behavior during marital conflict and immunological down-regulation. *Psychosomatic Medicine*, 55, 395–409.

Kingsbury, S. M. (1933). *Records of The Virginia Company of London*, vol. III. Washington, DC: US Gov. Printing Office.

Kinsey, A. C., Pomeroy, W. B. & Martin, C. E. (1948). *Sexual Behaviour in the Human Male*. Philadelphia: W. B. Saunders Co.

Kinsey, A. C., Pomeroy, W. B., Martin, C. E. & Gebhard, P. H. (1953). *Sexual Behaviour in the Human Female*. Philadelphia: W. B. Saunders Co.

Kirkwood, J. (1987). Animals at home – pets as pests: a review. *Journal of the Royal Society of Medicine*, 80, 97–100.

Kittredge, G. L. (1958). *Witchcraft in Old and New England*. New York: Russell & Russell.

Kleimann, D. G. (1967). Some aspects of social behaviour in the Canidae. *American Zoologist*, 7, 279–88.

Klein, R. (1995). The power of pets: America's misplaced obsession. *The New Republic*, 10 July, 18–23.

Kynge, J. (1986). Dogs barking up right tree in China. *Los Angeles Times*, 20 July.

Labarge, M. W. (1980). *A Baronial Household of the Thirteenth Century*. Brighton: Harvester Press.

Lago, D., Delaney, M., Miller, M. & Grill, C. (1989). Companion animals, attitudes towards pets, and health outcomes among the elderly: a long-term follow-up. *Anthrozoös*, 3, 25–34.

Larson, R., Csikszentmihalyi, M. & Graef, R. (1982). Time alone in daily experience: loneliness or renewal. In *Loneliness: A Source Book of Current Theory and Research*, eds. L. A. Peplau & D. Perlman, pp. 40–53. New York: John Wiley & Sons.

Laudenslager, M. L., Ryan, S. M., Drugan, R. C., Hyson, R. L. & Maier, S. F. (1983). Coping and immunosuppression: inescapable but not escapable shock suppresses lymphocyte secretion. *Science*, 221, 568–70.

Laughlin, W. S. (1968). Hunting: an integrating biobehaviour system and its evolutionary importance. In *Man the Hunter*, ed. R. B. Lee & I. DeVore, pp. 304–20. Chicago: Aldine Press.

Lawrence, E. A. (1989). Neoteny in American perceptions of animals. In *Perceptions of Animals in American Culture*, ed. R. J. Hoage, pp. 57–76. Washington, DC: Smithsonian Institution Press.

Leach, E. (1964). Anthropological aspects of language: animal categories and verbal abuse. In *New Directions in the Study of Language*, ed. E. H. Lenneberg, pp. 23–63. Cambridge, MA: MIT Press.

Leakey, R. E. & Lewin, R. (1977). *Origins*. London: Macdonald & Jane's.

Lee, D. (1983). Pet-therapy – helping patients through troubled times. *California Veterinarian*, 37, 24–5.

Lee, R. (1976). *The Pet Dog: Interactive Correlates of a Man–Animal Relationship*. University of Hull: Unpublished Report, Department of Psychology.

Lee, R. B. (1969). !Kung Bushmen subsistence: an input–output analysis. In *Environment and Cultural Behaviour*, ed. A. Vayda, pp. 47–79. Garden City: Natural History Press.

Lee, R. B. & DeVore, I. (1968). Problems in the study of hunter–gatherers. In *Man the Hunter*, eds. R. B. Lee & I. DeVore, pp. 3–12. Chicago: Aldine Press.

Leopold, A. (1968). *A Sand County Almanac and Sketches Here and There*. London: Oxford University Press.

Lesser, A. (1968). Warfare and the state. In *War: the Anthropology of Armed Conflict and Aggression*, ed. M. Friedman, M. Harris & R. Murphy, pp. 92–6. Garden City, NY: Natural History Press.

Lévi-Strauss, C. (1966). *The Savage Mind*. Chicago University Press.

Levine, M. (1971). Prehistoric art and ideology. In *Man in Adaptation: the Institutional Framework*, ed. Y. A. Cohen, pp. 424–34. Chicago: Aldine Press.

Levine, S. (1983). Coping: an overview. In *Biological and Psychological Basis of Psychosomatic Disease*, ed. H. Ursin & R. Murison, pp. 15–26. Oxford: Pergamon Press.

Levinson, B. M. (1962). The dog as co-therapist. *Mental Hygiene*, 46, 59–65.

(1964). Pets: a special technique in child psychotherapy. *Mental Hygiene*, 48, 243–8.

(1969). *Pet-oriented Child Psychotherapy*. Springfield, IL: Charles Thomas.

(1972). *Pets and Human Development*. Springfield, IL: Charles Thomas.

(1975). Pets and environment. In *Pet Animals and Society*, ed. R. S. Anderson, pp. 8–18. London: Baillière Tindall.

(1978). Pets and personality development. *Psychological Reports*, 42, 1031–8.

(1980). The child and his pet: a world of non-verbal communication. In *Ethology and Nonverbal Communication in Mental Health*, eds. S. A. Corson & E. O'Leary Corson, pp. 63–81. Oxford: Pergamon Press.

Linton, R. (1936). *The Study of Man: An Introduction*. New York: Appleton-Century-Crofts.

Linzey, A. (1987). *Christianity and the Rights of Animals*. New York: Crossroad.

Lockwood, R. (1983). The influence of animals on social perception. In *New Perspectives on Our Lives with Companion Animals*, eds. A. H. Katcher & A. M. Beck, pp. 64–71. Philadelphia: University of Pennsylvania Press.

(1989). Anthropomorphism is not a four letter word. In *Perceptions of Animals in American Culture*, ed. R. J. Hoage, pp. 41–56. Washington, DC: Smithsonian Institution Press.

(1995). The ethology and epidemiology of canine aggression. In *The Domestic Dog: Its Evolution, Behaviour & Interactions with People*, ed. J. A. Serpell, pp. 131–8. Cambridge University Press.

London, J. (1903). *The Call of the Wild*. London: Heinemann.

Lopez, B. H. (1978). *Of Wolves and Men*. New York: Charles Scribner's Sons.

Lorenz, K. (1943). Die angeborenen Formen möglicher Erfahrung. *Zeitschrift für Tierpsychologie*, 5, 235–409.

(1954). *Man Meets Dog*. London: Methuen.

Lovell, B. (1979). *In the Centre of Immensities*. London: Hutchinson.

Lumholtz, C. (1889). *Among Cannibals*. London: John Murray.

Luomala, K. (1960). The native dog in the Polynesian system of values. In *Culture in History*, ed. S. Diamond, pp. 190–240. New York: Columbia University Press.

Lynch, J. J. (1977). *The Broken Heart: The Medical Consequences of Loneliness*. New York: Basic Books.

Lynch, J. J. & McCarthy, J. F. (1966). Social responding in dogs: effect of person. *Conditional Reflex*, 1, 81.

Maccoby, H. (1982). *The Sacred Executioner*. London: Thames & Hudson.

MacKenzie, J. M. (1988). *The Empire of Nature: Hunting, Conservation and British Imperialism*. Manchester University Press.

Mader, B., Hart, L. A. & Bergin, B. (1989). Social acknowledgements for children with disabilities: effects of service dogs. *Child Development*, 60, 1529–34.

Maehle, A.-H. (1994). Cruelty and kindness to the 'brute creation': stability and change in the ethics of the man–animal relationship, 1600–1850. In *Animals & Human Society: Changing Perspectives*, eds. A. Manning & J. A. Serpell, pp. 81–105. London: Routledge.

Malarkey, W. B., Kiecolt-Glaser, J. K., Pearl, D. & Glaser, R. (1993). Hostile behavior during marital conflict alters pituitary and adrenal hormones. *Psychosomatic Medicine*, 56, 41–51.

Malek, J. (1993) *The Cat in Ancient Egypt*. London: British Museum Press.

Martel, F. L., Nevison, C. M., Rayment, F. D., Simpson, M. J. & Keverne, E. B. (1993). Opioid receptor blockade reduces maternal affect and social grooming in rhesus monkeys. *Psychoneuroendocrinology*, 18, 307–21.

Martin, C. (1978). *Keepers of the Game: Indian–Animal Relationships and the Fur Trade*. Berkeley, CA: University of California Press.

Martin, P. (in press). *The Sickening Mind: Brain, Behaviour, Immunity & Disease*. London: HarperCollins.

Martin, P. S. (1967). Prehistoric overkill. In *Pleistocene Extinctions: The Search for a Cause*, eds. P. S. Martin & H. E. Wright, pp. 75–120. New Haven, CT: Yale University Press.

Martin, W. R. (1977). General problems of drug-abuse and drug dependence. In *Drug Addiction 1: Handbook of Experimental Pharmacology*, vol. XLV, no. 1, ed. W. R. Martin, pp. 3–35. Berlin: Springer-Verlag.

Martinez, R. L. & Kidd, A. H. (1980). Two personality characteristics in adult pet-owners. *Psychological Reports*, 47, 381.

Marvin, G. (1988). *Bullfight*. Oxford: Blackwell.

Mason, J. (1985). Brave new farm? In *In Defence of Animals*, ed. P. Singer, pp. 89–107. Oxford: Basil Blackwell.

Mason, J. (1993). *An Unnatural Order: Uncovering the Roots of Domination of Nature and Each Other*. New York: Simon & Schuster.

Mathes, E. W. & Deuger, D. I. (1982). Jealousy, a creation of human culture? *Psychological Reports*, 51, 251–354.

McCaul, K. D. & Malott, J. M. (1984). Distraction and coping with pain. *Psychological Bulletin*, 95, 516–33.

McCulloch, M. J. (1983). Animal-facilitated therapy: overview and future direction. In *New Perspectives on Our Lives with Companion Animals*, eds. A. H. Katcher & A. M. Beck, pp. 410–26. Philadelphia: University of Pennsylvania Press.

McFarland, D. (1981). *The Oxford Companion to Animal Behaviour*. Oxford University Press.

McFarland, D. & McFarland, J. (1969). *An Introduction to the Study of Animal Behaviour*. Oxford: Basil Blackwell.

McNeill, W. H. (1983). *The Pursuit of Power*. Oxford: Basil Blackwell.

McNiell Taylor, L. (1983). *Living with Loss*. London: Fontana.

Mech, L. D. (1970). *The Wolf: The Ecology and Behavior of an Endangered Species*. Garden City, NY: Doubleday.

Meggit, M. J. (1965). The association between Australian Aborigines and dingoes. In *Man, Culture and Animals*, ed. A. Leeds & A. P. Vayda, pp. 17–18. Washington, DC: American Association for the Advancement of Science.

Mehrabian, A. (1971). Nonverbal betrayal of feeling. *Journal of Experimental Research in Personality*, 5, 64–73.

Menninger, K. A. (1951). Totemic aspects of contemporary attitudes toward animals. In *Psychoanalysis and Culture*, eds. G. B. Wilbur & W. Muensterberger, pp. 42–74. New York: John Wiley & Sons.

Merlen, R. H. A. (1971). *De Canibus: Dog and Hound in Antiquity*. London: J. A. Allen.

Messent, P. R. (1983). Social facilitation of contact with other people by pet dogs. In *New Perspectives on Our Lives with Companion Animals*, eds. A. H. Katcher & A. M. Beck, pp. 37–46. Philadelphia: University of Pennsylvania Press.

Messent, P. R. & Horsfield, S. (1985). Pet population and the pet–owner bond. In *The Human–Pet Relationship*, pp. 7–17. Vienna: IEMT – Institute for Interdisciplinary Research on the Human–Pet Relationship.

Messent, P. R. & Serpell, J. A. (1981). A historical and biological view of the pet–owner bond. In *Interrelations between People and their Pets*, ed. B. Fogle, pp. 5–22. Springfield, IL: Charles Thomas.

Midgley, M. (1983). *Animals and Why They Matter*. London: Penguin Books.

Mighetto, L. (1991). *Wild Animals and American Environmental Ethics*. Tucson: University of Arizona Press.

Mills, S. (1984). Big bad wolf? *New Scientist*, 1432, 28–31.

Mitchell, G. (1972). Looking behaviour in the rhesus monkey. *Journal of Phenomenological Psychology*, 3, 53–67.

Moelk, M. (1979). The development of friendly approach behaviour in the cat: a study of kitten–mother relations and the cognitive development of the kitten from birth to eight weeks. *Advances in the Study of Animal Behaviour*, 10, 164–224.

Morris, B. (1995). Woodland and village: reflections on the 'animal estate' in rural Malawi. *Journal of the Royal Anthropological Institute (N.S.)*, 1, 301–15.

Mowatt, F. (1963). *Never Cry Wolf*. Toronto: McClelland & Stuart.

Mugford, R. A. (1980). The social significance of pet-ownership. In *Ethology and Nonverbal Communication in Mental Health*, eds. S. A. Corson & E. O'Leary Corson, pp. 111–22. Oxford: Pergamon Press.

(1981). Problem dogs and problem owners: the behaviour specialist as an adjunct to veterinary practice. In *Interrelations between People and Pets*, ed. B. Fogle, pp. 295–315. Springfield: Charles C. Thomas.

(1983). The social skills of dogs as an indicator of animal awareness. In *Self-awareness in Domesticated Animals*, eds. D. G. H. Wood-Gush, M. Dawkins & R. Ewbank, pp. 40–4. Potter's Bar: Universities Federation for Animal Welfare.

(1995). Canine behavioural therapy. In *The Domestic Dog: Its Evolution, Behaviour & Interactions with People*, ed. J. A. Serpell, pp. 139–52. Cambridge University Press.

Murdock, G. P. (1968). The current status of the world's hunting and gathering peoples. In *Man the Hunter*, eds. R. B. Lee & I. DeVore, p. 13. Chicago: Aldine Press.

Nash, R. (1982). *Wilderness and the American Mind*. New Haven, CT: Yale University Press.

Natoli, E. (1984). Personal communication.

Nelson, R. K. (1986). A conservation ethic and environment: the Koyukon of Alaska. In *Resource Managers: North American and Australian Hunter–Gatherers*, eds. N. N. Williams & E. S. Hunn, pp. 211–28. Canberra: Institute of Aboriginal Studies.

Nerem, R. M., Levesque, M. J. & Cornhill, J. F. (1980). Social environment as a factor in diet-induced atherosclerosis. *Science*, 208, 1475–6.

Norman, G. (1985). Princess's father to sell family pictures. *The Times*, 14 May.

North, R. (1983). *The Animals Report*. London: Penguin Books.

O'Callaghan, M.-L. (1983). Pet issue becomes dog's dinner. *Guardian*, 13 October.

Oldroyd, D. R. (1980). *Darwinian Impacts: An Introduction to the Darwinian Revolution*. Milton Keynes: Open University Press.

Orwell, G. (1947). *The English People*. London: Collins.
Ory, M. G. & Goldberg, E. L. (1983). Pet possession and life satisfaction in elderly women. In *New Perspectives on Our Lives with Companion Animals*, eds. A. H. Katcher & A. M. Beck, pp. 303–17. Philadelphia: University of Pennsylvania Press.
Osgood, C. (1975). *The Chinese: The Study of a Hong Kong Community*. Tucson: University of Arizona Press.
Ouida. (1891). Dogs and their affections. *North American Review*, 153, 317.
Panksepp, J., Herman, B. H., Vilberg, T., Bisop, P. & DeEskinazi, F. G. (1980). Endogenous opioids and social behaviour, *Neuroscience and Biobehavioural Reviews*, 4, 473–87.
Papashvily, G. & Papashvily, H. (1954). *Dogs and People*. Philadelphia: J. B. Lippincott.
Papousek, M. & Papousek, M. (1977). Mothering and the cognitive head-start: psychobiological implications. In *Studies in Mother–Infant Interaction*, ed. H. R. Schaffer, pp. 63–85. New York & London: Academic Press.
Passmore, J. (1983). The man who had to eat his dog to survive. *Daily Mail*, 29 September.
Paul, E. S. & Serpell, J. A. (1993). Childhood pet keeping and humane attitudes in young adulthood. *Animal Welfare*, 2, 321–37.
Penny, N. B. (1976). Dead dogs and Englishmen. *The Connoisseur*, August, pp. 298–303.
Pepys, S. (1970–6). *The Diary of Samuel Pepys*, vols. I, IV & IX, eds. R. Latham & W. Matthews. London: G. Bell & Sons.
Perlman, D. & Peplau, L. A. (1981). Toward a social psychology of loneliness. In *Personal Relationships in Disorder: Personal Relationships*, vol. III, ed. S. Duck & R. Gilmour, pp. 31–56. New York & London: Academic Press.
PFI (1994). *PFI Fact Sheet 1994*. Washington, DC: Pet Food Institute.
PFMA (Pet Food Manufacturers' Association) (1995). *Profile*.
Phillips, B. (1982). What dogs are wearing. *The Observer*, 1 August.
Pitcairn, T. K. & Eibl-Eibesfeldt, I. (1976). Concerning the evolution of nonverbal communication in man. In *Communicative Behaviour and Evolution*, eds. M. E. Hahn & E. C. Simmel, pp. 81–113. New York & London: Academic Press.
Plumb, J. H. (1975). Pet Power. *Horizon*, 17, 4047.
Pond, W. G. (1983). Modern pork production. *Scientific American*, 248, 78–87.
Poresky, R. H., Hendrix, C., Mosier, J. E. & Samuelson, M. L. (1988). Young children's companion animal bonding and adults' pet attitudes: a retrospective study. *Psychological Reports*, 62, 419–25.

PRCA (Professional Rodeo Cowboys Association) (1994). *Humane Facts*. Colorado Springs: PRCA.

Price, E. O. (1984). Behavioural aspects of animal domestication. *Quarterly Review of Biology*, 59, 1–32.

Prioleau, L., Murdock, M. & Brody, N. (1983). An analysis of psychotherapy versus placebo studies. *The Behavioural and Brain Sciences*, 6, 275–310.

Pritchard, W. R. (1988). *Future Directions for Veterinary Medicine*. Durham, NC: Pew National Veterinary Education Program.

Purchas, S. (1906). *Purchas his Pilgrimes*, vol. XIX. Glasgow: James MacLehose & Sons.

Reed, C. A. (1954). Animal domestication in the prehistoric Near East. *Science*, 130, 1629–39.

Rice, S., Brown, L. & Caldwell, H. S. (1973). Animals and psychotherapy: a survey. *Journal of Community Psychology*, 1, 323–6.

Ritchie, C. I. A. (1981). *The British Dog: Its History from Earliest Times*. London: Robert Hale.

Ritvo, H. (1986). Our pets, our pork chops. *New York Times Book Review*, 26 October.

(1987). *The Animal Estate: The English and Other Creatures in the Victorian Age*. Cambridge, MA: Harvard University Press.

(1988). The emergence of modern pet-keeping. In *Animals and People Sharing the World*, ed. A. N. Rowan, pp. 13–31. Hanover, NH: University Press of New England.

Robb, S. S. & Stegman, C. E. (1983). Companion animals and elderly people: a challenge for evaluators of social support. *Gerontologist*, 23, 277–82.

Roberts, J. (1985). *The Triumph of the West*. London: BBC Publications.

Rosen, B. (1969). *Witchcraft*. London: Edward Arnold.

Ross, S. B. (1983). The therapeutic use of animals with the handicapped. *International Child Welfare Review*, 56, 26–39.

Roth, W. E. (1970). *An Introductory Study of the Arts, Crafts and Customs of the Guiana Indians*. New York: Johnson Reprint Corporation.

Rothschild, M. (1985). *The Relationship between Animals and Man*. Oxford: Romanes Lecture.

Rowan, A. N. (1995). Pet ownership and use of superstores on the rise. *Anthrozoös*, 8, 111.

Rowland, B. (1972). *Blind Beasts: Chaucer's Animal World*. Kent, OH: Kent State University Press.

Rowland, R. (1990). Fantasticall and devilishe persons: European witch-beliefs in comparative perspective. In *Early Modern European Witchcraft: Centres and Peripheries*, eds. B. Ankarloo & G. Henningsen, pp. 161–90. Oxford: Clarendon Press.

RSPCA (1984). The RSPCA in 1983. *News from the RSPCA*, 6 March.

Russel, W. M. S. (1956). On misunderstanding animals. *Universities Federation for Animal Welfare Courier*, 12, 19–35.

Rynearson, E. K. (1978). Humans and pets and attachment. *British Journal of Psychiatry*, 133, 550–5.

Sahlins, M. (1976). *Culture and Practical Reason*. Chicago University Press.

Salmon, I. M. & Salmon, P. W. (1982). *A Dog in Residence: A Companion Animal Study Undertaken at the Caulfield Geriatric Hospital*. Melbourne: Joint Advisory Committee on Pets in Society.

Salmon, P. W. & Salmon, I. (1983). Who owns who? Psychological research into the human–pet bond in Australia. In *New Perspectives on Our Lives with Companion Animals*, eds. A. H. Katcher & A. M. Beck, pp. 244–65. Philadelphia: University of Pennsylvania Press.

Savishinsky, J. S. (1983). Pet ideas: the domestication of animals, human behaviour, and human emotions. In *New Perspectives on Our Lives with Companion Animals*, eds. A. H. Katcher & A. M. Beck, pp. 112–31. Philadelphia: Pennsylvania University Press.

Sax, B. (1993). Holocaust images and other powerful ambiguities in the debate on animal experimentation: further thoughts. *Anthrozoös*, 6, 108–14.

Schachter, S. (1959). *The Psychology of Affiliation*. London: Tavistock Publications.

Schenkel, R. (1967). Submission: its features and functions in the wolf and dog. *American Zoologist*, 7, 319–30.

Schleifer, H. (1985). Images of death and life: food animal production and the vegetarian option. In *In Defence of Animals*, ed. P. Singer, pp. 63–73. Oxford: Basil Blackwell.

Schmitt, J.-C. (1983). *The Holy Greyhound: Guinefort, Healer of Children since the 13th Century*, trans. M. Thom. Cambridge University Press.

Schroer, J. (1994). Doggone! Animal owners are big spenders. *USA Today*, 2 March.

Sebkova, J. (1977). *Anxiety Levels as Affected by the Presence of a Dog*. Unpublished thesis, University of Lancaster.

Selye, H. (1957). *The Stress of Life*. London: Longman's, Green & Co.

Serpell, J. A. (1981). Childhood pets and their influence on adults' attitudes. *Psychological Reports*, 49, 651–4.

(1983a). What have we got against pets? *New Scientist*, 1379, 80–4.

(1983b). The personality of the dog and its influence on the pet–owner bond. In *New Perspectives on Our Lives with Companion Animals*, eds. A. H. Katcher & A. M. Beck, pp. 57–63. Philadelphia: University of Pennsylvania Press.

(1983c). *Research on Human–Animal Relationships: An Annotated Bibliography*, unpublished MS, International Symposium on the Human–Pet Relationship, Austrian Academy of Science, Vienna, 27 October.

(1985). Best friend or worst enemy: cross-cultural variation in attitudes to the domestic dog. In *The Human–Pet Relationship*, pp. 112–16. Vienna: IEMT – Institute for Interdisciplinary Research on the Human–Pet Relationship.

(1988a). Pet-keeping in non-Western societies: some popular misconceptions. In *Animals and People Sharing the World*, ed. A. N. Rowan, pp. 33–52. Hanover, NH: University Press of New England.

(1988b). The domestication and history of the cat. In *The Domestic Cat: The Biology of Its Behaviour*, eds. D. S. Turner & P. P. G. Bateson, pp. 151–8. Cambridge University Press.

(1989a). Pet-keeping and animal domestication: a reappraisal. In *The Walking Larder: Patterns of Domestication, Pastoralism, and Predation*, ed. J. Clutton-Brock, pp. 10–21. London: Unwin Hyman.

(1989b). Humans, animals, and the limits of friendship. In *The Dialectics of Friendship*, eds. R. Porter & S. Tomaselli, pp. 111–29. London: Routledge.

(1989c). Attitudes to animals. In *The Status of Animals: Ethics, Education & Welfare*, eds. D. Paterson & M. Palmer, pp. 162–6. Wallingford, Oxon: CAB. International.

(1990a). All the King's horses. *Anthrozoös*, 3, 223–6.

(1990b). Evidence for long term effects of pet ownership on human health. In *Waltham Symposium 20: Pets, Benefits and Practice*, ed. I. H. Burger, pp. 1–7. London: BVA Publications.

(1991). Beneficial effects of pet ownership on some aspects of human health and behaviour. *Journal of the Royal Society of Medicine*, 84, 717–20.

(1992). Animal protection and environmentalism: the background. In *Animal Welfare and the Environment*, ed. R. D. Ryder, pp. 27–39. London: Duckworth.

(1995). From paragon to pariah: some reflections on human attitudes to dogs. In *The Domestic Dog: Its Evolution, Behaviour & Interactions with People*, ed. J. A. Serpell, pp. 245–56. Cambridge University Press.

Serpell, J. A. & Jagoe, J. A. (1995). Early experience and the development of behaviour. In *The Domestic Dog: Its Evolution, Behaviour & Interactions with People*, ed. J. A. Serpell, pp. 79–102. Cambridge University Press.

Serpell, J. A. & Paul, E. S. (1994). Pets and the development of positive attitudes to animals. In *Animals & Human Society: Changing Perspectives*, eds. A. Manning & J. A. Serpell, pp. 127–44. London: Routledge.

Seyfarth, R. M. & Cheney, D. L. (1984). Grooming, alliances, and reciprocal altruism in vervet monkeys. *Nature*, 308, 541–3.

Sharefkin, B. D. & Ruchlis, K. (1974). Anthropomorphism in the lower grades. *Science and Children*, 11(6), 37–40.

Shell, M. (1988). The family pet. *Representations*, 15, 121–53.

Sheppard, R. Z. (1986) Pet theories and pet peeves. *Time*, 18 August, p. 49.

Sherick, I. (1981). The significance of pets for children: illustrated by a latency-age girl's use of pets in her analysis. *Psychoanalytic Studies of Children*, 36, 193–215.

Siegel, J. M. (1990). Stressful life events and use of physician services among the elderly: the moderating role of pet ownership. *Journal of Personality and Social Psychology*, 58, 1081–6.

Signoret, J. P., Baldwin, B. A, Frazer, D. & Hafez, E. S. E. (1975). The behaviour of swine. In *The Behaviour of Domestic Animals*, ed. E. S. E. Hafez, pp. 295–329. London: Baillière Tindall.

Simon, L. J. (1984). The pet trap: negative effects of pet-ownership on families. In *The Pet Connection*, eds. R. K. Anderson, B. L. Hart & L. A. Hart, pp. 226–40. Minneapolis: Center to Study Human–Animal Relationships and Environments, University of Minnesota.

Singer, M. (1968). Pygmies and their dogs: a note on culturally constituted defence mechanisms. *Ethos*, 6, 270–9.

Singer, P. (1984). *Animal Liberation*. Wellingborough: Thorson's Publishers.

Sklar, L. S. & Anisman, H. (1981). Stress and cancer. *Psychological Bulletin*, 89, 369–406.

Smith, R. (1984). Mother weeps for dog that ate her baby. *The Sun*, 18 March.

Smith, S. L. (1983). Interactions between pet dog and family members: an ethological study. In *New Perspectives on Our Lives with Companion Animals*, eds. A. H. Katcher & A. M. Beck, pp. 29–36. Philadelphia: University of Pennsylvania Press.

Sorabji, R. (1993). *Animal Minds & Human Morals: The Origins of the Western Debate*. Ithaca, NY: Cornell University Press.

Soulsby, L. & Serpell, J. A. (1988). *Companion Animals in Society: Council for Science & Society Report*. Oxford University Press.

Speck, F. G. (1935). *Naskapi*. Norman: University of Oklahoma Press.

Speke, J. H. (1863). *Journal of the Discovery of the Source of the Nile.* London: W. Blackwood.

Spencer, C. (1993). *The Heretic's Feast: A History of Vegetarianism.* London: Fourth Estate.

Sprenger, J. & Kramer, A. (1968). *Malleus Maleficarum*, trans. M. Summers. London: Folio Society.

Stallones, L., Marx, M. B., Garrity, T. F. & Johnson, T. P. (1990). Pet ownership and attachment in relation to the health of U.S. adults, 21–64 years of age. *Anthrozoös*, 4, 100–12.

Stein, M., Miller, A. H. & Trestman, R. L. (1991). Depression and the immune system. In *Psychoneuroimmunology*, 2nd edition, eds. R. Ader, D. L. Felten & N. Cohen, pp. 897–930. New York: Academic Press.

Sternglanz, S. H., Gray, J. L. & Murakami, M. (1977). Adult preferences for infantile facial features: an ethological approach. *Animal Behaviour*, 25, 108–15.

Studman, C. J. (1983). *The Dog Today: The Problems and the Solutions.* Universities Federation for Animal Welfare, Internal Report.

Syme, S. L. (1974). Behavioral factors associated with the etiology of physical disease. *American Journal of Public Health*, 64, 1043–5.

Szasz, K. (1968). *Petishism: Pet Cults of the Western World.* London: Hutchinson.

Tambiah, S. J. (1969). Animals are good to think and good to prohibit. *Ethnology*, 8, 452–3.

Tannenbaum, J. (1995). *Veterinary Ethics,* 2nd edn. St Louis, MO: Mosby-Year Book Inc.

ten Bensel, R. W. (1984). Historical perspectives of human values for animals and vulnerable people. In *The Pet Connection*, eds. R. K. Anderson, B. L. Hart & L. A. Hart, pp. 2–14. Minneapolis: CENSHARE, University of Minnesota.

Tester, K. (1991). *Animals & Society: The Humanity of Animal Rights.* London: Routledge.

Thomas, K. (1971). *Religion and the Decline of Magic.* London: Penguin Books.

(1983). *Man and the Natural World: Changing Attitudes in England 1500–1800.* London: Allen Lane.

Thorpe, W. H. (1972). The comparison of vocal communication in animals and man. In *Non-verbal Communication*, ed. R. A. Hinde, pp. 27–47. Cambridge University Press.

Toynbee, J. M. C. (1973). *Animals in Roman Life and Art.* London: Thames & Hudson.

Tuan, Yi-Fu. (1984). *Dominance and Affection: The Making of Pets.* New Haven, CT: Yale University Press.

Tucker, N. (1989). Animals in children's literature. In *The Status of Animals: Ethics, Education & Welfare*, eds. D. Paterson & M. Palmer, pp. 167–72. Wallingford: CAB International.

Turnbull, C. (1976). *The Forest People*. London: Pan Books.

UNEP (United Nations Environment Programme) (1989). *Environmental Data Report*. Oxford: Blackwell.

Vialles, N. (1994). *Animal to Edible*. Cambridge University Press.

Vilhjalmson, R. (1993). Life stress, social support and clinical depression: a reanalysis of the literature. *Social Science Medicine*, 37, 331–42.

Voith, V. L. (1981). Attachment between people and their pets: behavior problems of pets that arise from the relationship between pets and people. In *Interrelations between People and Pets*, ed. B. Fogle, pp. 271–94. Springfield, IL: Charles Thomas.

——— (1984). Human/animal relationships. In *Nutrition and Behaviour in Dogs and Cats*, ed. R. S. Anderson, pp. 147–56. Oxford: Pergamon Press.

Walker, C. (1984). Where white steak is unkosher pork. *The Times*, 17 December.

Watson, N. L. & Weinstein, M. (1993). Pet ownership in relation to depression, anxiety, and anger in working women. *Anthrozoös*, 4, 135–8.

Watts, D. (1985). Touching tribute to canine loyalty. *The Times*, 28 February.

WCMC (World Conservation Monitoring Centre) (1994). *Biodiversity Data Sourcebook*. Cambridge: WCMC Publications.

Weir, J. (1992). The Sweetwater rattlesnake round-up: a case study in environmental ethics. *Conservation Biology*, 6, 116–27.

Weiss, R. S. (1972). The provision of social relationships. In *Doing Unto Others*, ed. Z. Rubin, pp. 17–26. Englewood Cliffs, NJ: Prentice-Hall.

Wenzel, G. (1991). *Animal Rights, Human Rights: Ecology, Economy and Ideology in the Canadian Arctic*. London: Belhaven Press.

White, L. (1967). The historical roots of our ecologic crisis. *Science*, 155, 1203–7.

Wilbert, J. (1972). *Survivors of Eldorado: Four Indian Cultures of South America*. New York: Praeger.

Wilson, E. O. (1975). *Sociobiology*. Cambridge, MA: Belknap Press.

Windeler, R. (1983). A dog's life. *Family Weekly* (CA), 27 March.

Winkler, A., Fairnie, H., Gericevich, F. & Long, M. (1989). The impact of a resident dog on an institution for the elderly: effects on perceptions and social interactions. *The Gerontologist*, 29, 216–23.

Wolfensohn, S. (1981). The things we do to dogs. *New Scientist*, 1253,
    404–7.
Woods, S. M. (1965). Psychotherapy and the patient's pet. *Current
    Psychiatric Therapy*, 5, 119–21.
Woosey, B. (1983). Milady buys £8,000 jet trip for dog. *The Sun*,
    1 November.
Worth, D. & Beck, A. M. (1981). Multiple ownership of animals in
    New York city. *Transactions & Studies of the College of Physicians of
    Philadelphia*, 3, 280–300.
Wyllie, I. (1981). *The Cuckoo*. London: Batsford.
Yost, A. (1991). Twisting tales of potbellied piggies. *Washington Post*,
    25 July.
Zahn-Waxler, C., Hollenbeck, B. & Radke-Yarrow, M. (1985). The
    origins of empathy and altruism. In *Advances in Animal Welfare
    Science 1984–85*, eds. M. W. Fox & L. D. Mickley, pp. 21–39.
    Washington, DC: Humane Society of the United States.
Zeuner, F. E. (1963). *A History of Domesticated Animals*. London:
    Hutchinson.

Index